Illuminate Publishing

G000253149

Eduqas
Chemistry
A Level Year 1 & AS

Study and Revision Guide

Peter Blake
Elfed Charles
Kathryn Foster

Published in 2016 by Illuminate Publishing Ltd, P.O Box 1160, Cheltenham, Gloucestershire GL50 9RW

Orders: Please visit www.illuminatepublishing.com
or email sales@illuminatepublishing.com

© Peter Blake, Elfed Charles, Kathryn Foster

The moral rights of the authors have been asserted.

All rights reserved. No part of this book may be reprinted, reproduced or utilised in any form or by any electronic, mechanical, or other means, now known or hereafter invented, including photocopying and recording, or in any information storage and retrieval system, without permission in writing from the publishers.

British Library Cataloguing in Publication Data

A catalogue record for this book is available from the British Library

ISBN 978-1-908682-68-0

Printed by 4edge Ltd, Hockley, Essex

02.16

The publisher's policy is to use papers that are natural, renewable and recyclable products made from wood grown in sustainable forests. The logging and manufacturing processes are expected to conform to the environmental regulations of the country of origin.

Every effort has been made to contact copyright holders of material reproduced in this book. If notified, the publishers will be pleased to rectify any errors or omissions at the earliest opportunity.

This material has been endorsed by Eduqas and offers high quality support for the delivery of Eduqas qualifications. While this material has been through a Eduqas quality assurance process, all responsibility for the content remains with the publisher.

Eduqas examination questions are reproduced by permission from Eduqas

Editor: Geoff Tuttle
Design: Nigel Harriss
Layout: EMC Design Ltd, Bedford

Acknowledgements

We are very grateful to the team at Illuminate Publishing for their professionalism, support and guidance throughout this project. It has been a pleasure to work so closely with them.

The author and publisher wish to thank:

Judith Bonello for her thorough review of the book and expert insights and observations.

Contents

How to use this book

As experienced senior examiners for the Specification we have written this study guide to help you be aware of what is required, and structured the content to guide you through to success in the AS Chemistry examination, or any exam that you might sit that covers the equivalent content of the first year of the full A Level.

The content and structure of this book can be used to support the Eduqas AS Chemistry specification OR the first year of the full two-year Eduqas A Level Chemistry specification too.

If you are a student following the latter course using this book, you will find that the AS specification Sections C1.1 to C1.7 match to the A Level specification content Sections C1.1 to C1.6.

However, the AS Sections C1.7 and C2.1 to C2.8 map to the Eduqas A Level Sections C2.1 to C3.5 as follows:

AS   C1.7 C2.1 C2.2 C2.3 C2.4 C2.5 C2.6 C2.7 C2.8
A Level C2.1 C2.2 C2.3 C2.4 C3.1 C3.2 C3.3 C3.4 C3.5

Knowledge and Understanding

The **first section** of the book covers the key knowledge and understanding required for the examination and provides notes for each of the two examination theory papers:

Component 1 The Language of Chemistry, Structure of Matter and Simple Reactions

Component 2 Energy, Rate and the Chemistry of Carbon Compounds

In addition, we have tried to give you additional pointers so that you can develop your work:

- Questions may be based on any term in the Specification so these terms are defined and highlighted.
- There are 'Quickfire' questions designed to test your knowledge and understanding of the material.
- 'Pointers' pick out things that may be useful in answering questions.
- 'Grade Boost' inserts point out key ways in which candidates can impress the examiners by their knowledge and understanding.
- 'Extra' comprises a type of Quickfire question that increases in difficulty.
- There is a comprehensive set of candidate answers to questions in all sections along with marking, analysis and explanation by the examiners of these answers.

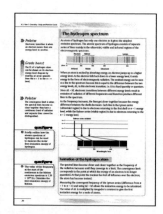

Exam Practice and Technique

The **second section** of the book covers the key skills for examination success and offers you examples based on real-life responses to examination questions. First, you will be guided into an understanding of how the examination system works, and then offered clues to success.

A variety of structured and essay questions are provided in this section. Each essay includes the marking points expected followed by actual samples of candidates' responses. A variety of structured questions is also provided, together with typical responses and comments. They offer a guide as to the standard that is required, and the commentary will explain why the responses gained the marks that they did.

Most importantly, we advise that you should take responsibility for your own learning and not rely on your teachers to give you notes or tell you how to gain the grades that you require. You should look for additional notes to support your study of AS Chemistry.

We advise that you look at the Eduqas website www.eduqas.co.uk. In particular, you need to be aware of the Specification. Look for specimen examination papers and mark schemes. You may find past papers useful as well.

Good luck with your revision.

Peter Blake, Elfed Charles and Kathryn Foster

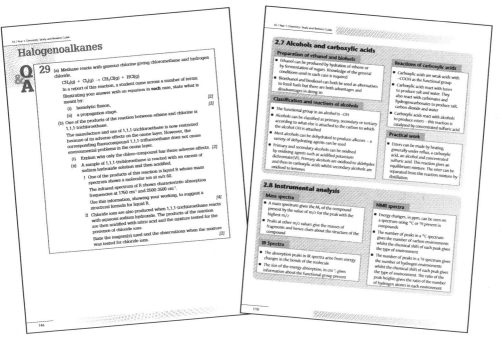

C1 Knowledge and Understanding

The Language of Chemistry, Structure of Matter and Simple Reactions

This unit begins with an introduction to chemical language followed by some important fundamental ideas about atoms and the use of the mole concept in calculations.

The usefulness of materials depends on their properties that in turn depend on their internal structure and bonding. By understanding the relationship between these, chemists can design new useful materials. The types of forces between particles are studied along with several types of solid structures to show how these influence properties. The building blocks of materials are the elements and the relationship of their properties to their position in the Periodic Table is illustrated by a study of the elements of the s-block and Group 7.

The key principles governing the position of equilibrium between reactants and products are considered and applied to the important field of acid-base reactions.

Revised it!

Basic notes *Good grasp* *Fully revised*

1.1 Formulae and equations

The ability to represent reactions using balanced chemical (and ionic) equations is essential in chemistry. Using chemical formulae enables equations to be written. Oxidation number can be used to express the combining power of elements to form compounds.

→ **p8–10** → ☐ ☐ ☐

1.2 Basic ideas about atoms

Atoms have an internal structure comprising protons, neutrons and electrons. Some atoms are unstable and the nucleus splits up to form smaller particles – this is called radioactive decay. Radioactive emission can have adverse consequences or can be used beneficially. Evidence for the arrangement of electrons in an atom is given by ionisation energies, and emission and absorption spectra.

→ **p11–20** → ☐ ☐ ☐

1.3 Chemical calculations

Masses of atoms are expressed relative to the carbon-12 isotope. Relative atomic masses are determined by mass spectrometry. The mole can be used to calculate the masses of solids, concentration of solutions or volumes of gases reacting or being formed. The concentration of an unknown solution can be found using an acid-base titration and percentage error in measurements can be calculated.

→ **p21–33** → ☐ ☐ ☐

1.4 Bonding

Atoms bond together to form molecules by electrical forces either through covalency with the sharing of electron pairs or by the transfer of electrons from one atom to another to form ionic bonds. The type of bond depends on the difference in electronegativity between the atoms. Bonding between molecules is weak (van der Waals forces), although hydrogen bonds are stronger. Molecular shapes are governed by the number of electron pairs around a central atom following the VSEPR theory.

p34–40

1.5 Solid structures

Solids may be either giant or simple molecules; ionic and metallic solids always being giants comprising millions of units, and covalent structures either giants as in diamond or simple molecules such as I_2 that are held in the solid by weak van der Waals forces. Physical properties of the solids, such as melting temperature and electrical conductivity, relate to underlying structure of the solid.

p41–42

1.6 Periodic Table

Arranging the elements according to their electronic structures as in the Periodic Table is a very powerful means of rationalising and understanding chemical behaviour. Systematic trends in both physical and chemical properties emerge, in particular the typical valencies and redox properties of the groups. Detailed examples are shown for the s-block elements – reducing agents and metals – and the halogens of Group 7 – oxidising agents and non-metals.

p43–46

1.7 Simple equilibria and acid-base reactions

Many reactions are reversible and when the forward and reverse reactions occur at the same rate the reaction has reached equilibrium. The position of equilibrium changes if the conditions of a reaction change. The ratio of concentrations of products to reactants can be expressed mathematically using the equilibrium constant. All acids are proton donors, some are strong, others weak. The strength of acids is measured using the pH scale. Acid-base titrations provide opportunities to link practical work to chemical calculations.

p47–54

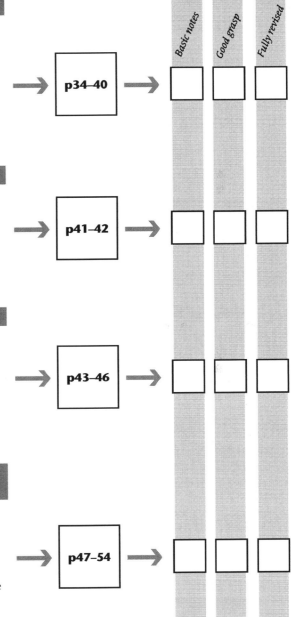

Basic notes

Good grasp

Fully revised

1.1 Formulae and equations

Formulae of compounds and ions

Grade boost

Learn the formulae of the common acids and common gases.

The formula of a compound is a set of symbols and numbers. The symbols say what elements are present and the numbers give the ratio of the numbers of atoms of the different elements in the compound.

E.g.

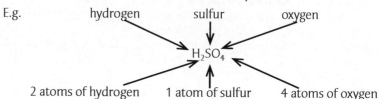

① What is the formula of:
   a) Hydrochloric acid?
   b) Sulfuric acid?
   c) Ammonia?
   d) Methane?

Many compounds consist of molecules in which the atoms are bonded covalently. To show multiple molecules you write the appropriate number in front of the formula. This number multiplies everything after it.

E.g. in $3HNO_3$ atoms of hydrogen = $3 \times 1 = 3$
atoms of nitrogen = $3 \times 1 = 3$
atoms of oxygen = $3 \times 3 = 9$

Many compounds do not consist of molecules but consist of ions and they form through ionic bonding. For an ionic compound the total number of positive charges must equal the total number of negative charges in one formula unit of the compound.

quickfire

② How many atoms of each element are present in $2CH_3CO_2H$?

The formula for ionic compounds can be calculated by following these steps:

1 Write the symbols of the ions in the compound.

2 Balance the ions so that the total of the positive ions and negative ions adds to zero. (The compound itself must be neutral.)

quickfire

③ How many atoms of oxygen are present in $Ca(HCO_3)_2$?

3 Write the formula without the charges and put the number of ions of each element as a small number following and below the element symbol. (Use brackets to denote more than one compound ion.)

Example 1 Sodium oxide

1 Sodium is in group 1, oxygen is in group 6, therefore the ions are Na^+ and O^{2-}.

2 Two Na^+ ions are needed to balance the charge on one O^{2-} ion to make the total charge zero ($+1 +1 -2 = 0$).

3 Formula is Na_2O (the '1' does not need to be included).

Pointer

You do not need to learn the formula for ionic compounds.

Example 2 Calcium hydroxide

1 Calcium is in group 2, hydroxide is a compound ion, the ions are Ca^{2+} and OH^-.

Grade boost

Learn the formula for Group 1, 2, 6 and 7 ions.

2 Two OH^- ions are needed to balance the charge on one Ca^{2+} ion to make the total charge zero ($-1 -1 +2 = 0$).

3 Formula is $Ca(OH)_2$ (note the use of a bracket around the OH before adding the 2).

quickfire

④ Write the formula for:
   a) Sodium carbonate
   b) Barium sulfate
   c) Ammonium sulfate.

Oxidation numbers

A method of expressing the combining power of elements is oxidation number. The oxidation number of an element is the number of electrons that need to be added to (or taken away from) an element to make it neutral.

E.g., the ion magnesium, Mg^{2+}, forms when magnesium loses two electrons, therefore it needs the addition of two electrons to make a neutral atom and has the oxidation number +2. The chloride ion, Cl^-, needs to lose an electron to make a neutral atom; therefore it has the oxidation number −1.

The following table gives the rules for assigning oxidation numbers:

Rule	Example
The oxidation number of an uncombined element is zero.	Metallic copper, Cu: oxidation number 0 Oxygen gas, O_2: oxidation number 0.
The sum of the oxidation numbers in a compound is zero. In an ion, the sum equals the overall charge.	In CO_2 the sum of the oxidation numbers of carbon and oxygen is 0. In NO_3^- the sum of the oxidation numbers of nitrogen and oxygen is −1.
In compounds, the oxidation numbers of Group 1 metals is +1 and Group 2 metals is +2.	In $MgBr_2$ the oxidation number of magnesium is +2 (oxidation number of each bromine is −1).
The oxidation number of oxygen is −2 in compounds except with fluorine or in peroxides (and superoxides).	In SO_2 the oxidation number of each oxygen is −2 (oxidation number of sulfur is +4). In H_2O_2 the oxidation number of oxygen is −1 (oxidation number of hydrogen is +1).
The oxidation number of hydrogen is +1 in compounds except in metal hydrides.	In HCl the oxidation number of hydrogen is +1 (oxidation number of chlorine is −1). In NaH the oxidation number of hydrogen is −1 (oxidation number of sodium is +1).
In chemical species with atoms of more than one element, the most electronegative element is given the negative oxidation number.	In CCl_4, chlorine is more electronegative than carbon, so oxidation number of each chlorine is −1 (oxidation number of carbon is +4).

Example

What is the oxidation number of manganese in MnO_4^-?

The oxidation number of each oxygen is −2, so the total for oxygen is $-2 \times 4 = -8$

The total charge on the ion is −1 = ox. no. Mn + total ox. no. O

Therefore −1 = ox. no. Mn + (−8)

Ox. no. Mn = +7

>> Pointer

You do not need to know the definition of oxidation number.

Grade boost

Learn the rules for assigning oxidation numbers.

quickpire

⑤ What is the oxidation number of:
a) Carbon in CH_4
b) Phosphorus in Na_3PO_4
c) Nitrogen in KNO_2?

Chemical and ionic equations

Grade boost

Remember that all the elements in the mnemonic HOFBrINCl exist as diatomic molecules.

Pointer

The following state symbols can be included in an equation: (s) for solid, (l) for liquid, (g) for gas and (aq) for a solution.

quickfire

⑥ Write a balanced chemical equation for the reaction between:

a) sodium carbonate and hydrochloric acid

b) nitrogen dioxide (NO_2) and water to give nitric oxide (NO) and nitric acid.

quickfire

⑦ Write an ionic equation, including state symbols for the following reactions:

a) adding magnesium to hydrochloric acid

b) adding a solution of lead(II) nitrate to a solution of potassium iodide to form the yellow precipitate lead(II) iodide.

Chemical equations are written to sum up what happens in a chemical reaction using chemical formulae. However, since atoms are neither created nor destroyed in a chemical reaction there must be the same number of atoms of each element on each side of the chemical equation. To balance a chemical equation, all you can do is to multiply a formula in the equation by putting a number in front of the formula.

E.g. hydrogen burns in oxygen to form water

$$H_2 + O_2 \longrightarrow H_2O$$

Counting the number of atoms on each side gives:

L.H.S. Hydrogen 2 atoms, oxygen 2 atoms

R.H.S. Hydrogen 2 atoms, oxygen 1 atom

To balance the equation we need 2 atoms of oxygen on the R.H.S.

A simple solution would be to write $H_2 + O_2 \longrightarrow H_2O_2$

This is obviously incorrect since H_2O_2 is hydrogen peroxide not water. You cannot change a formula, you can only put a number in front of the formula to multiply by it. Since 2 atoms of oxygen are needed on the R.H.S. multiply the water by 2.

$$H_2 + O_2 \longrightarrow 2H_2O$$

However, we now have:

L.H.S. Hydrogen 2 atoms, oxygen 2 atoms

R.H.S. Hydrogen 4 atoms, oxygen 2 atoms

So multiply the hydrogen on the L.H.S. by 2

$$2H_2 + O_2 \longrightarrow 2H_2O$$

and the equation is balanced.

Many reactions involve ions in solutions. However, in these reactions not all of the ions take part in any chemical change. An ionic equation may help to show what is happening. It provides a shorter equation which focuses attention on the changes taking place. Any ions that do not change during a reaction are left out of an ionic equation. These ions are known as spectator ions.

Ionic equations are frequently used for displacement and precipitaton reactions.

E.g. barium sulfate is insoluble in water. When a solution of barium chloride is added to a solution of sodium sulfate a white precipitate forms. Write an ionic equation, including state symbols, for the reaction.

The chemical equation for the reaction is:

$$BaCl_2(aq) + Na_2SO_4(aq) \longrightarrow BaSO_4(s) + 2NaCl(aq)$$

Writing out all the ions gives:

$$Ba^{2+}(aq) + 2Cl^-(aq) + 2Na^+(aq) + SO_4^{2-}(aq) \longrightarrow$$
$$BaSO_4(s) + 2Na^+(aq) + 2Cl^-(aq)$$

The $Na^+(aq)$ ions and the $Cl^-(aq)$ ions do not change during the reaction. They are spectator ions and are omitted, giving the ionic equation:

$$Ba^{2+}(aq) + SO_4^{2-}(aq) \longrightarrow BaSO_4(s)$$

1.2 Basic ideas about atoms

Atomic structure

Atoms consist of a nucleus made up of positively charged protons and uncharged neutrons surrounded by shells containing negatively charged electrons moving non-stop. Almost all of the atom's mass is in the nucleus. An atom has the same number of protons and electrons.

Elements

Each element has its own **atomic number**.

We often incorporate the atomic number and the **mass number** into the symbol of an element, e.g. the full symbol of fluorine is $^{19}_{9}F$.

9 is the atomic number and 19 is the mass number.

Most elements exist naturally as a mixture of atoms which are only different in their mass numbers. These atoms are called **isotopes**. For example, naturally occurring chlorine consists of two isotopes, one having a mass number of 35 and one having a mass number of 37 or $^{35}_{17}Cl$ and $^{37}_{17}Cl$.

Ions

Positive **ions** (or cations) form when an atom loses one or more electrons

e.g. $K \longrightarrow K^+ + e^-$

Negative **ions** (or anions) form when an atom gains one or more electrons

e.g. $F + e^- \longrightarrow F^-$

In both examples the number of protons has not changed.

>> *Pointer*

Although atomic structure is not specifically mentioned in the specification, prior GCSE knowledge about it is expected.

Key Terms

Atomic number is the number of protons in the nucleus of an atom.

Mass number is the number of protons + the number of neutrons in the nucleus of an atom.

Isotopes are atoms having the same number of protons but different numbers of neutrons.

Grade boost

Don't forget, in any atom:
The atomic number =
the number of protons
The number of protons =
the number of electrons
The number of neutrons =
the mass number – the atomic number

a. Give the number of protons, neutrons and electrons in the two main isotopes of zinc, Zn-64 and Zn-66.

b. State the number of protons and electrons in (i) $^{16}O^{2-}$ (ii) $^{207}Pb^{2+}$.

c. State the difference, if any, between the chemical properties of Zn-64 and Zn-66, giving a reason for your answer.

Radioactivity

Types of radioactive emission and their behaviour

Radiation	Nature	Effect of electric field	Effect of magnetic field	Penetrating power
α **particles**	clusters of 2 protons and 2 neutrons	attracted to negative plate	deflected in certain direction	least penetrating stopped by a piece of paper
β **particles**	electrons	attracted to positive plate	deflected in opposite direction	stopped by a thin sheet of metal (e.g. 0.5 cm of aluminium)
γ **rays**	high energy electromagnetic radiation	no effect	no effect	most penetrating may take more than 2 cm of lead to stop them

Key Terms

α **particles** – cluster of 2 protons and 2 neutrons, therefore positively charged.

β **particles** – fast moving electrons, therefore negatively charged.

γ **rays** – high energy electromagnetic radiation, therefore no charge.

Grade boost

β particles can be considered as being formed when a neutron changes into a proton i.e.
$$^1_0n \rightarrow ^1_1p + _{-1}\beta$$

Pointer

A positron is a type of β particle. It has the same mass as an electron and a numerically equal but positive charge.

quickfire

① Why are victims of radioactive contamination buried in lead coffins?

quickfire

② Give the mass number and symbol of the isotope formed when ^{18}F decays by positron emission.

Effect on mass number and atomic number

α and β particle emissions result in the formation of a new nucleus with a new atomic number therefore the product is a different element.

When an element emits an α particle its mass number decreases by 4 and its atomic number decreases by 2.

$$^{238}_{92}U \rightarrow ^{234}_{90}Th + ^4_2\alpha$$

The product is two places to the left in the Periodic Table.

When an element emits a β particle its mass number is unchanged and its atomic number increases by 1.

$$^{14}_{6}C \rightarrow ^{14}_{7}N + _{-1}\beta$$

The product is one place to the right in the Periodic Table.

In the process of electron capture, the mass number is unchanged and the atomic number decreases by 1.

$$^{59}_{28}Ni + e^- \rightarrow ^{59}_{27}Co$$

The product is one place to the left in the Periodic Table.

In the process of positron emission, the mass number is unchanged and the atomic number decreases by 1.

$$^{12}_{6}C \rightarrow ^{12}_{5}B + \beta^+$$

The product is one place to the left in the Periodic Table.

Half-life

The nuclei in different radioactive substances decay at different rates. The time taken for one half of all nuclei in a radioisotope to decay is known as its **half-life**. The half-life depends only on which isotope is decaying, not on the quantity present.

Calculations involving half-life can be set to find:

- The time taken for the radioactivity of a sample to fall to a certain fraction of its initial value.
- The mass of a radioactive isotope remaining after a certain length of time given the initial mass.
- The half-life of a radioisotope given the time taken for a sample to fall to a certain fraction of its initial value.

Example 1

Tritium has a half-life of 13 years. How many radioactive nuclei will be left after 52 years from an original sample which contained 16 million nuclei?

52 years = 4 half-lives

16 million $\xrightarrow{13}$ 8 million $\xrightarrow{13}$ 4 million $\xrightarrow{13}$ 2 million $\xrightarrow{13}$ 1 million

Example 2

An isotope of phosphorus, ^{32}P, is radioactive and its radioactivity falls to 1/8th of its initial value in 42.9 days. Calculate the half-life of ^{32}P.

1 $\longrightarrow$ 1/2 $\longrightarrow$ 1/4 $\longrightarrow$ 1/8 is 3 half-lives

Half-life = $\dfrac{42.9}{3}$ = 14.3 days

Consequences for living cells

Radioactive emissions are potentially harmful. However, we all receive some radiation from the normal background radiation that occurs everywhere. Workers in industries where they are exposed to radiation from radioactive isotopes are carefully monitored to ensure that they do not receive more radiation than is allowed under internationally agreed limits.

High energy radioactive emissions break chemical bonds in the cell molecules giving rise to changes in DNA which can cause mutations and the formation of cancerous cells at lower doses, or cell death at higher doses.

When α-emitting isotopes are ingested they are far more dangerous than an equivalent activity of β emitting or γ emitting isotopes but fortunately α particles from outside cannot penetrate the skin.

Key Term

Half-life is the time taken for half the atoms in a radioisotope to decay or the time taken for the radioactivity of a radioisotope to fall to half its initial value.

>> *Pointer*

The greater the half-life of a radioactive isotope, the greater the concern, since the radioactivity of the isotope exists for a longer time.

quickfire

③ ^{60}Co used in radiotherapy has a half-life of 5.3 years. Calculate how long it would take for the activity of the isotope to decay to 1/16th of its original value.

a. Outline why radioactivity may be a health hazard.

b. Why is α radiation the most harmful if ingested but least harmful outside the body?

Beneficial uses of radioactivity

Below are some examples. Candidates should be able to give an example in each area.

quickfire

④ Radioactive iodine, ^{131}I is used in medicine as a tracer. Give another use of a named radioactive isotope apart from in medicine.

Medicine

- Cobalt-60 in radiotherapy for the treatment of cancer. The high energy of γ radiation is used to kill cancer cells and prevent the malignant tumour from developing.
- Iodine-131 for patients with defective thyroid glands. The iodine-131 acts as a tracer to study the uptake of iodine in the gland.
- Technetium-99m is the most commonly used medical radioisotope. It is used as a tracer, normally to label a molecule which is preferentially taken up by the tissue to be studied.

Radio-dating

- Carbon-14 (half-life 5570 years) is used to calculate the age of plant and animal remains. All living organisms absorb carbon, which includes a small proportion of the radioactive carbon-14. When an organism dies there is no more absorption of carbon-14 and that which is already present decays. The rate of decay decreases over the years and the activity that remains can be used to calculate the age of organisms.
- Potassium-40 (half-life 1300 million years) is used to estimate the geological age of rocks. Potassium-40 can change into argon-40 by the nucleus gaining an inner electron. Measuring the ratio of potassium-40 to argon-40 in a rock gives an estimate of its age.

Analysis

- Dilution analysis. The use of isotopically labelled substances to find the mass of a substance in a mixture. This is useful when a component of a complex mixture can be isolated from the mixture in the pure state but cannot be extracted quantitatively.

quickfire

⑤ Explain why β emitters are used to measure the thickness of metal strips.

- Monitoring the thickness of metal strips or foil. The metal is placed between two rollers to get the right thickness. A radioactive source (a β emitter) is mounted on one side of the metal with a detector on the other. If the amount of radiation reaching the detector increases, the detector operates a mechanism for moving the rollers apart and vice versa.

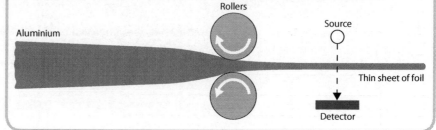

Electronic structure

Electrons within atoms occupy fixed energy levels or shells. Shells are numbered 1, 2, 3, 4, etc. These numbers are known as principal quantum numbers, n. The lower the value of n, the closer the shell to the nucleus and the lower the energy level.

Electron subshells or orbitals

In a shell there are regions of space around the nucleus where there is a high probability of finding an electron of a given energy. These regions are called **atomic orbitals**. Orbitals of the same type are grouped together in a subshell. Each orbital can contain two electrons. Along with charge, electrons have a property called spin, this reduces the effect of repulsion. In order for two electrons, both of which have a negative charge, to exist in the same orbital they must have opposite spins. There are four different types of orbital: s, p, d, and f. Orbitals of the same type are grouped together as a subshell.

An s orbital is spherical and can contain two electrons.

  The boundary represents an area where the electron spends 90% of its existence.

A p orbital is made up of three dumb-bell shaped lobes mutually at right angles. They are shown separated below:

p_x orbital p_y orbital p_z orbital

Since each p orbital can hold two electrons, a p subshell can hold 6 electrons in total.

There are five d orbitals; therefore there are a total of 10 electrons in a d subshell.

This is how the Periodic Table looks in terms of s, p and d electrons.

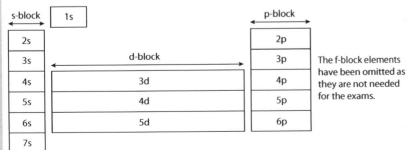

The f-block elements have been omitted as they are not needed for the exams.

Key Term

Atomic orbital is a region in an atom that can hold up to two electrons with opposite spins.

» Pointer

The s subshell can hold 2 electrons. The p subshell can hold 6 electrons. The d subshell can hold 10 electrons.

Grade boost

Sodium is classified as an s-block element because its outer electron is in an s orbital. Chlorine is classified as a p-block element because its outer electron is in a p orbital.

Key Term

Electronic configuration
is the arrangement of
electrons in an atom.

Grade boost

The 4s orbitals are filled
before the 3d orbitals.
The configurations for
chromium and copper are not
as expected, they both end
in 4s^1.

Pointer

You need to know the
electronic configuration for
the first 36 elements.

quickfire

⑥ (a) Write the electronic
configuration in terms
of subshells for a
copper atom.

(b) Use electrons in
boxes to write
the electronic
configuration of:

(i) an atom of silicon

(ii) an oxide ion, O^{2-}.

Filling shells and orbitals with electrons

The way in which electrons are arranged is called **electronic** structure or
configuration. This can be worked out using three basic rules:

1 Electrons fill atomic orbitals in order of increasing energy.

2 A maximum of two electrons can occupy any orbital each with opposite
spins.

3 Each orbital in a subshell will first fill with one electron before pairing
starts.

The most common way of representing electronic structure is to write the
shell number first, followed by the orbital letter and then the number of
electrons in the orbital which are written as a superscript.

For example, since nitrogen has:

2 electrons in the s orbital in the 1st shell

2 electrons in the s orbital in the 2nd shell

3 electrons in the p orbital in the 2nd shell,

its electronic configuration is $1s^2 2s^2 2p^3$.

Calcium has 20 electrons, its electronic configuration is : $1s^2 2s^2 2p^6 3s^2 3p^6 4s^2$.

Another way is to use 'electrons in boxes'. Each orbital is represented as a box
and the electrons as arrows in the boxes. The opposite spin of paired electrons
is shown by arrows facing up and down.

e.g. Nitrogen has 7 electrons

| ⇅ | ⇅ | ↑ | ↑ | ↑ |

1s   2s   2p

The electronic configuration of ions is presented in the same way as that of
atoms.

Positive ions form by the loss of electrons from the highest energy orbitals so
these ions have fewer electrons than the parent atom.

Negative ions form by adding electrons to the highest energy orbitals so these
ions have more electrons than the parent atom.

e.g. Na $1s^2 2s^2 2p^6 3s^1$ Na$^+$ $1s^2 2s^2 2p^6$

Cl $1s^2 2s^2 2p^6 3s^2 3p^5$ Cl$^-$ $1s^2 2s^2 2p^6 3s^2 3p^6$

Ionisation energies

The process of removing electrons from an atom is called ionisation. The process for the **first ionisation energy** (IE) of an element is summarised in the equation:

$$X(g) \longrightarrow X^+(g) + e^-$$

Electrons are held in their shells by their attraction to the positive nucleus, therefore the greater the attraction, the greater the ionisation energy. This attraction depends on three factors:

- **Nuclear charge** – the greater the nuclear charge, the greater the attractive force on the outer electron.
- **Electron shielding** – the more filled inner shells or subshells there are, the smaller the attractive force on the outer electron.
- **Distance of outer electron from nucleus** – the greater the distance, the smaller the attractive force on the outer electron.

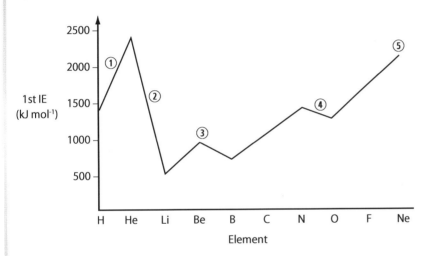

A plot of first IE against the first ten elements of the Periodic Table shows evidence of shells and subshells:

1. He > H since helium has a greater nuclear charge in the same subshell so little extra shielding.
2. He > Li since lithium's outer electron is in a new shell which has increased shielding and is further from the nucleus.
3. Be > B since boron's outer electron is in a new subshell of slightly higher energy level and is partly shielded by the 2s electrons.
4. N > O since the electron–electron repulsion between the two paired electrons in one p orbital in oxygen makes one of the electrons easier to remove. Nitrogen does not contain paired electrons in its p orbital.
5. He > Ne since neon's outer electron has increased shielding from inner electrons and is further from the nucleus.

Key Terms

The **molar first ionisation energy** of an element is the energy required to remove one mole of electrons from one mole of its gaseous atoms.

Electron shielding or **screening** is the repulsion between electrons in different shells. Inner shell electrons repel outer shell electrons.

>> *Pointer*
If the conditions for ionisation energy are 298 K and 1 atm then the process is known as the standard ionisation energy.

quickfire

⑦ State and explain the general trend in ionisation energy:
a) across a period
b) down a group.

Successive ionisation energies

Successive ionisation energies are a measure of the energy needed to remove each electron in turn until all the electrons are removed from an atom.

An element has as many ionisation energies as it has electrons. Sodium has eleven electrons and so has eleven successive ionisation energies.

For example, the third ionisation energy is a measure of how easily a 2^+ ion loses an electron to form a 3^+ ion. An equation to represent the third ionisation energy of sodium is:

$$Na^{2+}(g) \longrightarrow Na^{3+}(g) + e^-$$

Successive ionisation energies always increase because:

- There is a greater effective nuclear charge as the same number of protons are holding fewer and fewer electrons.
- As each electron is removed there is less electron–electron repulsion and each shell will be drawn in slightly closer to the nucleus.
- As the distance of each electron from the nucleus decreases, the nuclear attraction increases.

The graph below shows how the successive ionisation energies of sodium provide further evidence for the existence of different shells.

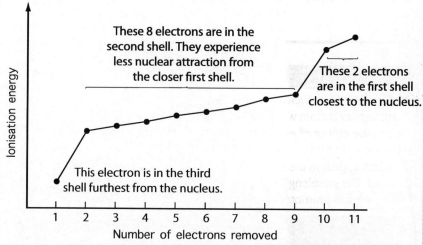

For sodium there is one electron on its own which is easiest to remove. Then there are eight more electrons which become successively more difficult to remove. Finally there are two electrons which are the most difficult to remove.

Notice the large increases in ionisation energy as the 2nd and 10th electrons are removed. If the electrons were all in the same shell, there would be no large rise or jump.

Grade boost

A large increase in successive ionisation energies shows that an electron has been removed from a new shell closer to the nucleus and gives the group to which the element belongs.
Li has a large energy jump between 1st and 2nd IE therefore it's in group 1.
Al has a large energy jump between 3rd and 4th IE therefore it's in group 3.

quickfire

⑧ Write an equation to represent the second ionisation energy of magnesium.

quickfire

⑨ The first four ionisation energies (in kJ mol^{-1}) for an element are: 590, 1150, 4940 and 6480.

State and explain to which group in the Periodic Table the element belongs.

Emission and absorption spectra

Light and electromagnetic radiation

Light is a form of electromagnetic radiation. The frequency and wavelength of light are related by the equation:

$$c = f\lambda \quad \text{(c is the speed of light)}$$

The frequency of electromagnetic radiation and energy (E) are connected by the equation:

$$E = hf \quad \text{(h is Planck's constant)}$$

Therefore, $f \propto E$, and if frequency increases energy increases.

$f \propto 1/\lambda$ and if frequency increases wavelength decreases.

The whole range of frequencies of electromagnetic radiation is called the electromagnetic spectrum. This diagram shows the ultraviolet, visible and infrared part of it.

						Frequency (Hz)
10^{11}	10^{12}	10^{13}	10^{14}	10^{15}	10^{16}	10^{17}

Far IR	IR	UV	Far UV

VISIBLE

Red	Orange	Yellow	Green	Blue	Violet
700	620	580	540	480	400

Wavelength (nm)

Grade boost

Since $f \propto E$ and $f \propto 1/\lambda$, the lower the wavelength, the higher the frequency and the greater the energy.

Pointer

Light is electromagnetic radiation in the range of wavelength corresponding to the visible region of the electromagnetic spectrum.

quickfire

(10) The first line in the visible emission spectrum for hydrogen has a wavelength of 656 nm. The visible emission spectrum of neon shows a prominent line at 585 nm. State and explain which line has the higher a) frequency b) energy.

quickfire

(11) State briefly the difference between absorption and emission spectra.

Absorption spectra

Light of all visible wavelengths is called white light. All atoms and molecules absorb light of certain wavelengths. Therefore, when white light is passed through the vapour of an element, certain wavelengths will be absorbed by the atoms and removed from the light. Looking through a spectrometer, black lines appear in the spectrum where light of some wavelengths has been absorbed. The wavelengths of these lines correspond to the energy taken in by the atoms to promote electrons from lower to higher energy levels.

Emission spectra

When atoms are given energy by heating or by an electrical field, electrons are promoted from a lower energy level to a higher one. When the source of energy is removed, the electrons fall from the higher energy level to a lower energy level and the energy lost is released as a packet of energy called a quantum of energy. This corresponds to electromagnetic radiation of a specific frequency. The observed spectrum consists of a number of coloured lines on a black background.

The hydrogen spectrum

Pointer

Electronic transition is when an electron moves from one energy level to another.

Grade boost

The IE of a hydrogen atom can be shown on its electron energy level diagram by drawing an arrow upwards from the $n = 1$ to the $n = \infty$ level.

Pointer

The convergence limit is when the spectral lines become so close together they have a continuous band of radiation and separate lines cannot be distinguished.

quickfire

⑫ Briefly outline how the atomic spectrum of hydrogen can be used to measure the molar first ionisation energy of hydrogen.

quickfire

⑬ The value of the frequency at the start of the continuum in the lithium emission spectrum is 1.30×10^{15} Hz. Calculate the first ionisation energy of lithium.

An atom of hydrogen has only one electron so it gives the simplest emission spectrum. The atomic spectrum of hydrogen consists of separate series of lines mainly in the ultraviolet, visible and infrared regions of the electromagnetic spectrum.

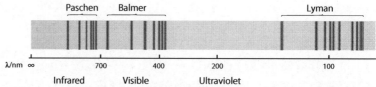

When an atom is excited by absorbing energy, an electron jumps up to a higher energy level. As the electron falls back down to a lower energy level, it emits energy in the form of electromagnetic radiation. The emitted energy can be seen as a line in the spectrum because this is equal to the difference between the two energy levels, ΔE, in this electronic transition, i.e. it is a fixed quantity or quantum.

Since $\Delta E = hf$, electronic transitions between different energy levels result in emission of radiation of different frequencies and therefore produce different lines in the spectrum.

As the frequency increases, the lines get closer together because the energy difference between the shells decreases. Each line in the Lyman series (ultraviolet region) is due to electrons returning to the first shell or $n = 1$ energy level, while the Balmer series (visible region) is due to electrons returning to the $n = 2$ energy level.

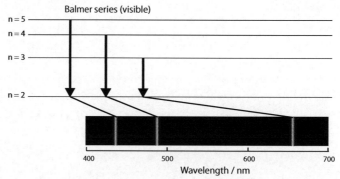

Ionisation of the hydrogen atom

The spectral lines become closer and closer together as the frequency of the radiation increases until they converge to a limit. The convergence limit corresponds to the point at which the energy of an electron is no longer quantised. At that point the nucleus has lost all influence over the electron; the atom has become ionised.

Measuring the convergent frequency of the Lyman series (difference from $n = 1$ to $n = \infty$) and using $\Delta E = hf$ allows the ionisation energy to be calculated. The value of ΔE is multiplied by Avogadro's constant to give the first ionisation energy for a mole of atoms.

1.3 Chemical calculations

Relative mass terms

Masses of atoms

The masses of atoms are too small to be used in calculations in chemical reactions, so instead the mass of an atom is expressed relative to a chosen standard atomic mass. The carbon-12 isotope is taken as the standard of reference.

Most elements exist naturally as two or more different isotopes. The mass of an element therefore depends on the relative abundance of all the isotopes present in the sample. In order to overcome this, chemists use an average mass of all the atoms and this is called the **relative atomic mass, A_r**.

Relative atomic mass has no units since it is one mass compared to another mass.

If we refer to the mass of a particular isotope then the term **relative isotopic mass** is used.

Masses of compounds

Since the formula of a compound shows the ratio in which the atoms combine, the idea of relative atomic mass can be extended to compounds and the term **relative formula mass, M_r**, is used.

E.g. the relative formula mass of copper(II) sulfate, $CuSO_4$ is:

$(1 \times 63.5) + (1 \times 32) + (4 \times 16) = 159.5$

Key Terms

Relative atomic mass is the average mass of one atom of the element relative to one-twelfth the mass of one atom of carbon-12.

Relative isotopic mass is the mass of an atom of an isotope relative to one-twelfth the mass of an atom of carbon-12.

Relative formula mass is the average mass of a molecule relative to one-twelfth the mass of an atom of carbon-12.

≫ Pointer

Relative formula mass used to be called relative molecular mass, but relative molecular mass really refers to compounds containing only molecules.

quickfire

① What is the relative formula mass, M_r, of:
(a) $Pb(NO_3)_2$
(b) $MgSO_4.7H_2O$

The mass spectrometer

>> *Pointer*

You do not need to be able to draw a diagram of a mass spectrometer but you could be required to label a diagram of one.

In order to calculate the average mass of an atom of an element, the mass of the isotopes of the element together with their relative abundances must be known. These values are found using a **mass spectrometer**.

The processes in a mass spectrometer may be summarised as:

- Vaporisation – sample is heated before it enters the spectrometer.
- Ionisation – a gaseous sample is bombarded with high energy electrons to form positive ions.
- Acceleration – an electric field accelerates the positive ions to high speed.
- Deflection – a magnetic field deflects the ions according to their mass/charge ratio. (Heavier ions are deflected less than light ones.)
- Detection – ions with the correct mass/charge ratio pass through a slit and are detected by an instrument such as an electrometer. The signal is then amplified and recorded.

The processes take place under high vacuum to prevent collision with air molecules.

a. State why the space inside a mass spectrometer is connected to a vacuum pump.

b. A sample of strontium showed three peaks. The first peak was at m/z 86 and had an abundance of 10%; the second peak was at m/z 87 and had an abundance of 7%; the third peak was at m/z 88 and had an abundance of 83%. Calculate the relative atomic mass to three significant figures.

c. Chlorine has a relative atomic mass of 35.5. It consists of two isotopes, one of which has an abundance of 75% and a mass of 35.0. Show that the mass of the second isotope is 37.0.

Calculating relative atomic masses

Below is the mass spectrum of magnesium

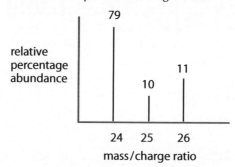

There are three peaks in the spectrum, so magnesium has three isotopes. The heights of the peaks give the relative abundances of the isotopes, which are given as percentages in the spectrum.

The relative atomic mass is a weighted average of the masses of all the atoms in the isotopic mixture, therefore:

$$\text{Relative atomic mass} = \frac{(79 \times 24) + (10 \times 25) + (11 \times 26)}{100} = 24.32$$

Other uses of mass spectrometry include:

- Identifying unknown compounds e.g. testing athletes for prohibited drugs.
- Identifying trace compounds in forensic science.
- Analysing molecules in space.

The mass spectrum of chlorine

Chlorine is made up of two isotopes ^{35}Cl and ^{37}Cl. However, chlorine gas consists of molecules not individual atoms and the mass spectrum of chlorine is:

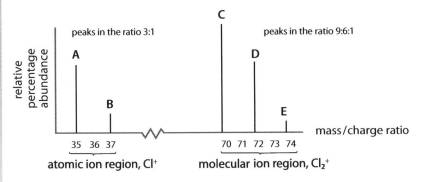

Mass spectrum of chlorine

When chlorine is passed into the ionisation chamber, an electron is knocked off the molecule to give a molecular ion, Cl_2^+. These ions won't be particularly stable, and some will fall apart to give a chlorine atom and a Cl^+ ion. (This is known as fragmentation.)

So peak A is caused by $^{35}Cl^+$ and peak B by $^{37}Cl^+$.

As the ^{35}Cl isotope is three times more common than the ^{37}Cl isotope, the heights of the peaks are in the ratio of 3:1.

In the molecular ion region think about the possible combinations of ^{35}Cl and ^{37}Cl atoms in a Cl_2^+ ion. Both atoms could be ^{35}Cl, both atoms could be ^{37}Cl, or you could have one of each sort.

So peak C (m/z 70) is due to $(^{35}Cl-^{35}Cl)^+$

Peak D (m/z 72) is due to $(^{35}Cl-^{37}Cl)^+$ or $(^{37}Cl-^{35}Cl)^+$

Peak E (m/z 74) is due to $(^{37}Cl-^{37}Cl)^+$

Since the probability of an atom being ^{35}Cl is $\frac{3}{4}$ and that of being ^{37}Cl is $\frac{1}{4}$, then

molecule	$^{35}Cl-^{35}Cl$	$^{35}Cl-^{37}Cl$ or $^{37}Cl-^{35}Cl$	$^{37}Cl-^{37}Cl$
probability	$\frac{3}{4} \times \frac{3}{4}$	$\frac{3}{4} \times \frac{1}{4}$ or $\frac{1}{4} \times \frac{3}{4}$	$\frac{1}{4} \times \frac{1}{4}$
	$\frac{9}{16}$	$\frac{6}{16}$	$\frac{1}{16}$

and ratio of peaks C:D:E is 9:6:1

>> **Pointer**

A molecular ion, M^+, is the positive ion formed in mass spectrometry when a molecule loses an electron, its mass gives the relative formula mass of the compound.

>> **Pointer**

For the chlorine spectrum you can't make any predictions about the relative heights of the lines at m/z 35/37 compared with those at 70/72/74. That depends on what proportion of the molecular ions break up into fragments.

quickfire

② In the mass spectrum of chlorine, explain why peaks due to chlorine atoms are present although chlorine gas contains only Cl_2 molecules.

>> **Pointer**

You will be expected to use information about molecular ions to deduce information about isotopes and vice versa for substances other than chlorine.

Key Terms

One mole is the amount of any substance that contains the same number of particles as there are atoms in exactly 12 g of carbon-12.

The Avogadro constant is the number of atoms per mole.

Molar mass is the mass of one mole of a substance.

Pointer

You do not have to recall the value of the Avogadro constant.

Grade boost

Learn the equations that connect amount of substance and mass for a solid:

$$n = \frac{m}{M} \quad \text{or} \quad m = nM \quad \text{or} \quad M = \frac{m}{n}$$

quickfire

③ a) Calculate the amount, in moles, of 1.86 g sodium hydroxide.

b) 0.020 moles of a compound weighed 1.48 g. Calculate the molar mass of the compound.

Amount of substance

In chemical reactions the atoms that make up the reactants rearrange to form the products. For all the reactants to change into products the correct quantity of each reactant must be used. Since atoms are too small to be counted individually, chemists count atoms by weighing a collection of them where the mass of a particular fixed number of atoms is known.

Again carbon-12 is chosen as the standard and the number of atoms in exactly 12 g of carbon-12 is called a **mole**. This is a large number, 6.02×10^{23}, and is called the **Avogadro constant, L**.

When using mole to describe the amount of substance it is important to state the particles to which it refers. One mole of oxygen atoms is different to one mole of oxygen molecules.

The mass per mole of an element or compound is called the **molar mass, M**. It has the same numerical value as A_r or M_r but has the unit $g\ mol^{-1}$.

The amount of substance and mass are linked by the equation:

$$\text{number of moles (n)} = \frac{\text{mass of substance (m)}}{\text{molar mass (M)}}$$

Example

If you need 0.20 moles of sodium carbonate, Na_2CO_3, what mass of the substance do you need to weigh out?

Molar mass of Na_2CO_3 = 106 $gmol^{-1}$

Mass of sample = $n \times M$ = 0.20×106 = 21.2 g

Empirical and molecular formulae

Empirical formula is the simplest formula showing the simplest whole number ratio of the amount of elements present.

Molecular formula shows the actual number of atoms of each element present in the molecule. It is a simple multiple of the empirical formula. Usually the relative formula mass is needed to determine the molecular formula.

Example

A compound of carbon, hydrogen and oxygen has a relative molecular mass of 180. The percentage composition by mass is C 40.0%; H 6.70%; O 53.3%. What is (a) the empirical formula and (b) the molecular formula?

(a)

	C	:	H	:	O
Molar ratio of atoms	40		6.7		53.3
	12		1.01		16
	= 3.33		6.63		3.33
Divide by smallest number	1		2		1
Empirical formula is		CH_2O			

(b) Mass of empirical formula = 12 + 2.02 + 16 = 30.02

Number of CH_2O units in a molecule $= \dfrac{180}{30.02} = 6$

Molecular formula is $C_6H_{12}O_6$

Some salts have water molecules incorporated into their structure. These are known as hydrated salts and the water is known as water of crystallisation. If we know the mass of the anhydrous salt and the mass of the water in the hydrated salt we can calculate the number of moles of water in the hydrated salt.

Example

When 6.04 g of hydrated calcium nitrate, $Ca(NO_3)_2.xH_2O$ were heated, 4.20 g of the anhydrous salt, $Ca(NO_3)_2$, remained. What is the value of x?

Mass of water in the hydrate = 6.04 − 4.20 = 1.84 g

Moles $Ca(NO_3)_2 = \dfrac{4.20}{164.1} = 0.0256$

Moles $H_2O = \dfrac{1.84}{18.02} = 0.102$

Mole ratio of	H_2O	:	$Ca(NO_3)_2$
	0.102	:	0.0256
Divide by smallest	3.98	:	1

Value of $x = 4$

Key Terms

Empirical formula is the simplest formula showing the simplest whole number ratio of the amount of elements present.

Molecular formula shows the actual number of atoms of each element present in the molecule. It is a simple multiple of the empirical formula.

Grade boost

When you divide the percentages by the relevant atomic masses, do not truncate the answers, e.g. 1.25 to 1. The figures provided in the question should give reasonably simple ratios for the empirical formula.

quickfire

 1.07 g of vanadium react with chlorine to form 2.56 g of vanadium chloride. Find the empirical formula of this compound.

» Pointer

Stoichiometry is the molar relationship between the amounts of reactants and products in a chemical reaction.

Calculating reacting masses

An equation tells us not only what substances react together but also what amounts of substances react together. The ratio between amounts in moles of reactants and products are called stoichiometric ratios (mole ratios). The number of moles of solids can be calculated from their masses, so if we are given the mass of reactants we can calculate the mass of products formed and vice versa.

Example

An oxide of lead, Pb_3O_4, used in anti-corrosive paints can be formed by oxidising lead(II) oxide, PbO, with oxygen.

$$6PbO + O_2 \longrightarrow 2Pb_3O_4$$

Calculate the mass of Pb_3O_4 that could be formed from 134 g of PbO.

Step 1   Change the amount, in moles, of PbO

$$n = \frac{m}{M} = \frac{134}{223} = 0.600$$

Step 2   Use the equation to calculate the amount, in moles, of Pb_3O_4 formed

6 mol PbO gives 2 mol Pb_3O_4

0.600 mol PbO gives 0.200 mol Pb_3O_4

Step 3   Calculate the mass of Pb_3O_4

$0.200 \times 685 = 137$ g

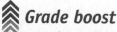

⑤ In the reaction
$6PbO + O_2 \longrightarrow 2Pb_3O_4$
calculate the volume of oxygen, at 0°C and 1 atm, that is needed to completely react with 2.23g of PbO.

» Pointer

Molar volume, v_m, is the volume per mole of a gas. There is no need to recall values for v_m.

Grade boost

Learn the equations that connect amount of substance and volume for a gas.

$$n = \frac{v}{v_m} \quad or \quad v = n \times v_m$$

Calculating volumes of gases

For reactions involving gases, it is more usual to consider the volumes of reactants and products rather than their masses. The numbers of moles of gases are calculated from their volumes using the molar gas volume, v_m.

Example

In the presence of a catalyst, potassium chlorate(V) decomposes on heating to give potassium chloride and oxygen according to the equation:

$$2KClO_3(s) \longrightarrow 2KCl(s) + 3O_2(g)$$

1.226 g of $KClO_3$ is heated until it is fully decomposed.

Calculate the volume of oxygen produced at 0°C and 1 atm.

(1 mole of oxygen occupies 22.4 dm^3 at 0°C and 1 atm).

Step 1   Calculate the amount, in mol, of 1.226 g $KClO_3$

$$n = \frac{m}{M} = \frac{1.226}{122.6} = 0.0100$$

Step 2   Calculate the amount of O_2 produced

2 mol $KClO_3$ gives 3 mol O_2

0.01 mol $KClO_3$ gives 0.015 mol O_2

Step 3   Calculate the volume of O_2 that is produced.

Volume $O_2 = n \times 22.4 = 0.015 \times 22.4 = 0.336$ dm^3 or 336 cm^3.

Volumes that gases occupy are dependent on the temperature and pressure used e.g. 1 mole of oxygen occupies 22.4 dm³ at 0°C and 1 atm but 24.0 dm³ at 25°C. To calculate the volume a gas would occupy at temperatures and pressures other than those at which it was actually measured we use the equation of state for an ideal gas:

$$\frac{P_1V_1}{T_1} = \frac{P_2V_2}{T_2}$$

where 1 represents the experimental conditions and 2 the new conditions.

The units for pressure and volume are not important as long as they are the same for both conditions. However, the units for temperature must be kelvins, K. (K = °C + 273)

Example

A sample of gas collected at 20°C and 1.00 atm pressure had a volume of 56.0 cm³. What is the volume at 50°C and 1 atm pressure?

$$\frac{P_1V_1}{T_1} = \frac{P_2V_2}{T_2}$$

$$\frac{1 \times 56}{293} = \frac{1 \times V_2}{323} \quad \text{(Temperature must be in kelvins)}$$

$$V_2 = \frac{1 \times 56 \times 323}{293} = 61.7 \text{ cm}^3$$

The ideal gas equation

The ideal gas equation was derived from gas laws and Avogadro's principle, and is written as PV = nRT.

In calculations involving the ideal gas equation SI units must be used, i.e.

Pressure must be in Pa (pascals) or Nm⁻² (Newtons per square metre)

Volume must be in m³

Temperature must be in K (kelvins)

Example

An ideal gas occupies a volume of 250 cm³ at 25°C and a pressure of 1.01×10^5 Nm⁻². Calculate the amount, in moles, of the gas in this sample.

$$PV = nRT$$

Therefore $n = \dfrac{PV}{RT}$

SI units must be used:

P = 1.01×10^5 Nm⁻²

V = 2.50×10^{-4} m³   (250 / 1 000 000)

T = 298 K   (25 + 273)

$$n = \frac{1.01 \times 10^5 \times 2.50 \times 10^{-4}}{8.31 \times 298}$$

$$n = 0.0102 \text{ mol}$$

Grade boost

To correct gas volumes due to changes in temperature or pressure use

$$\frac{P_1V_1}{T_1} = \frac{P_2V_2}{T_2}$$

remember to change temperature to kelvin.

quickfire

⑥ After 25.0 cm³ of a gas were collected at 70°C and 8.65×10^4 Nm⁻², conditions were changed to 25°C and 1.20×10^5 Nm⁻². What was the new volume?

Grade boost

Learn the ideal gas equation PV = nRT. You do not need to recall the value of R.

≫ Pointer

To change from °C to K add 273.
To change from cm³ to m³ divide by 10⁶ (1 000 000).

quickfire

⑦ Calculate the volume, in dm³ occupied by 0.65 mole of an ideal gas at 1.01×10^5 Pa and 20°C.

Grade boost

Learn the equations that connect amount of substance and concentration for a solution:

$$n = cv \quad or \quad c = \frac{n}{v} \quad or \quad v = \frac{n}{c}$$

Remember if v is given in cm^3 divide by 1000 to change it into dm^3.

quickfire

⑧ What mass of sodium carbonate is needed to prepare 250 cm^3 of a 0.400 mol dm^{-3} solution?

quickfire

⑨ The solubility of sodium chloride at 20°C is 36 $g/100cm^3$ water. What is the concentration in mol dm^{-3}?

Grade boost

Analysis of titration results follows a set pattern:
Calculate the moles of the solution for which the volume and concentration are given.
Use the mole ratio in the equation to find the moles of the other solution.
Change the moles into the answer required (concentration, molar mass, etc.).

Concentrations of solutions

The concentration of a solution measures how much of a dissolved substance is present per unit volume of a solution. The concentration of a solution can be stated in many ways but the most convenient way is to state the amount in moles of a solid present in 1 dm^3 of solution.

i.e. concentration (c) = $\dfrac{\text{amount of moles of solute (n)}}{\text{volume of solution (v)}}$

and the unit is mol dm^{-3}.

In the laboratory, volumes are normally measured in cm^3, so they have to be divided by 1000 to change them into dm^3.

Example

4.65 g of potassium nitrate, KNO_3, are dissolved in 200 cm^3 of water. What is the concentration of the solution in mol dm^{-3}?

$$\text{Moles } KNO_3 = \frac{4.65}{101.1} = 0.0460$$

$$\text{Concentration} = \frac{0.0460}{0.200} = 0.230 \text{ mol } dm^{-3}$$

Acid-base titration calculations

In acid-base titrations, since the reacting volumes of the solutions are known, if the concentration of one of the solutions is known and the stoichiometric ratios of the solutions are known the concentration of the unknown solution can be determined. If both concentrations are known, the stoichiometric ratio of the solutions can be determined.

Examples

1 20.0 cm^3 of sulfuric acid were exactly neutralised by 24.0 cm^3 of 0.950 mol dm^{-3} aqueous sodium hydroxide. Calculate the concentration of the acid.

$$H_2SO_4 + 2NaOH \longrightarrow Na_2SO_4 + 2H_2O$$

Step 1 Calculate the amount, in moles, of NaOH that reacted

$$n = c \times v = 0.95 \times 0.024 = 0.0228 \ (2.28 \times 10^{-2})$$

(Remember to divide 24 by 1000 to change it to dm^3)

Step 2 Use the equation to deduce the amount, in moles, of H_2SO_4 used

From the equation, 2 moles NaOH require 1 mole H_2SO_4

Therefore 0.0228 moles NaOH require 0.0114 moles H_2SO_4

Step 3 Calculate the concentration, in mol dm^{-3}, of H_2SO_4.

$$c = \frac{n}{v} = \frac{0.0114}{0.020} = 0.570 \text{ mol } dm^{-3}$$

2 A 500 cm³ solution of an acid (M_r = 90) was made by dissolving 2.81 g in water. On titration, 20.0 cm³ of this solution were completely neutralised by 25.0 cm³ of 0.100 mol dm⁻³ aqueous sodium hydroxide. Determine the stoichiometric ratio of acid:sodium hydroxide.

$$\text{Moles acid} = \frac{2.81}{90} = 0.0312$$

$$\text{Concentration} = \frac{0.0312}{0.500} = 0.0624 \text{ mol dm}^{-3}$$

Moles acid in 20.0 cm³ = $0.0624 \times 0.0200 = 1.248 \times 10^{-3}$

Moles sodium hydroxide = $0.100 \times 0.0250 = 2.50 \times 10^{-3}$

Moles acid	:	moles sodium hydroxide
1.248×10^{-3}	:	2.50×10^{-3}
1	:	2

3 This is an example of a back titration where the base is an insoluble salt.

An impure sample of calcium hydroxide of mass 0.716 g was added to 50.0 cm³ of 0.400 mol dm⁻³ hydrochloric acid solution. When the excess of acid was titrated against 0.225 mol dm⁻³ sodium hydroxide solution, 12.4 cm³ of sodium hydroxide solution were required to neutralise the acid. Calculate the percentage purity of the sample of calcium hydroxide.

Step 1 Calculate the amount, in moles, of NaOH used in the neutralisation, this will give the amount of moles of HCl unreacted in the original reaction.

The equation for the neutralisation reaction is

$$NaOH + HCl \longrightarrow NaCl + H_2O$$

Moles NaOH = $0.225 \times 0.0124 = 2.79 \times 10^{-3}$

Therefore moles HCl in excess = 2.79×10^{-3}

Step 2 Calculate the amount, in moles, of HCl that reacted with the impure $Ca(OH)_2$ sample.

Initial moles HCl = $0.400 \times 0.0500 = 2.00 \times 10^{-2}$

Moles HCl reacted with sample = $2.00 \times 10^{-2} - 2.79 \times 10^{-3} = 1.71 \times 10^{-2}$

Step 3 Calculate the amount of $Ca(OH)_2$ reacting with the HCl.

The equation for the reaction is

$$Ca(OH)_2 + 2HCl \longrightarrow CaCl_2 + 2H_2O$$

Mole ratio HCl : $Ca(OH)_2$ is 2 : 1

Therefore moles $Ca(OH)_2$ reacting = $1.71 \times 10^{-2} / 2 = 8.61 \times 10^{-3}$

Step 4 Calculate the percentage of $Ca(OH)_2$ in the sample.

Mass $Ca(OH)_2$ = $8.61 \times 10^{-3} \times 74.12 = 0.638$ g

$$\% \ Ca(OH)_2 \text{ in sample} = \frac{0.638}{0.716} \times 100 = 89.1 \ \%$$

extra

a. 20.0 cm³ of aqueous sodium hydroxide were exactly neutralised by 16.0 cm³ of 0.250 mol dm⁻³ aqueous hydrochloric acid. Calculate the concentration of the sodium hydroxide.

b. 4.76 g of washing soda, which is hydrated sodium carbonate, were dissolved in water and the solution made up to 250 cm³. A 25.0 cm³ portion of this solution required 33.2 cm³ of 0.100 mol dm⁻³ aqueous hydrochloric acids solution for neutralisation. Calculate the percentage by mass of sodium carbonate in washing soda.

c. A 0.400 g sample of impure ammonium chloride was warmed with an excess of sodium hydroxide solution. After the ammonia evolved had reacted with 25.0 cm³ of 0.150 mol dm⁻³ sulfuric acid, the excess of sulfuric acid required 11.40 cm³ of 0.100 mol dm⁻³ sodium hydroxide solution for neutralisation. Calculate the percentage of ammonium chloride in the original sample.

The reactions taking place are

$NH_4Cl + NaOH \longrightarrow$
$NH_3 + NaCl + H_2O$
$2NH_3 + H_2SO_4 \longrightarrow$
$(NH_4)_2SO_4$
$H_2SO_4 + 2NaOH \longrightarrow$
$Na_2SO_4 + H_2O$

4 This is an example of a double titration where a solution contains a mixture of two bases of different strengths.

25.0 cm³ of a solution containing sodium carbonate and sodium hydrogencarbonate needed 16.50 cm³ of a 0.200 mol dm⁻³ solution of hydrochloric acid to decolourise phenolphthalein. On addition of methyl orange, a further 28.50 cm³ of the acid were needed to turn the indicator orange.

The equations for the reactions taking place are:

$$CO_3^{2-} + H^+ \longrightarrow HCO_3^- \qquad \text{at phenolphthalein stage}$$

$$\left.\begin{array}{l} CO_3^{2-} + H^+ \longrightarrow HCO_3^- \\ HCO_3^- + H^+ \longrightarrow CO_2 + H_2O \end{array}\right\} \text{at methyl orange stage}$$

Calculate the concentrations of the sodium carbonate and sodium hydrogencarbonate in the solution.

The first stage relates to the amount of moles of carbonate only.

The second stage relates to the amount of moles of carbonate + hydrogencarbonate.

Moles acid in first stage = $0.200 \times 0.0165 = 3.30 \times 10^{-3}$

Therefore moles $CO_3^{2-} = 3.30 \times 10^{-3}$

Moles acid in second stage = $0.200 \times 0.0285 = 5.70 \times 10^{-3}$

Therefore total moles $CO_3^{2-} + HCO_3^- = 5.70 \times 10^{-3}$

and moles $HCO_3^- = 2.40 \times 10^{-3}$

Concentration $Na_2CO_3 = \dfrac{3.30 \times 10^{-3}}{0.025} = 0.132$ mol dm⁻³

Concentration $NaHCO_3 = \dfrac{2.40 \times 10^{-3}}{0.025} = 0.096$ mol dm⁻³

Atom economy and percentage yield

When a reaction occurs, the compounds formed, other than the product needed, are a waste. An indication of the efficiency of a reaction can be given as its **atom economy** or its **percentage yield**. The higher the atom economy, the more efficient the process.

Example

Benzenecarboxylic acid, C_6H_5COOH, can be formed from ethylbenzenecarboxylate, $C_6H_5COOC_2H_5$, according to the equation:

$$C_6H_5COOC_2H_5 + H_2O \rightarrow C_6H_5COOH + C_2H_5OH$$

If 1.95 g of acid were obtained from 3.00 g of ethylbenzenecarboxylate, calculate:

a) the percentage yield

b) the atom economy for the reaction

a) % yield = $\dfrac{\text{mass of product obtained}}{\text{maximum theoretical mass}} \times 100\%$

To calculate the maximum mass the 3 steps on page 26 must be followed:

Step 1 Calculate the amount, in mol, of 3.00 g $C_6H_5COOC_2H_5$

$$n = \frac{m}{M} = \frac{3.00}{150.1} = 0.0200$$

Step 2 Use the equation to calculate the amount, in mol, of C_6H_5COOH formed

1 mol $C_6H_5COOC_2H_5$ gives 1 mol C_6H_5COOH

0.02 mol $C_6H_5COOC_2H_5$ gives 0.02 mol C_6H_5COOH

Step 3 Calculate the mass of C_6H_5COOH

$$m = nM = 0.02 \times 122.06 = 2.44 \text{ g}$$

% yield $= \dfrac{1.95}{2.44} \times 100 = 79.9\%$

b) Atom economy = $\dfrac{\text{mass of required product}}{\text{total mass of reactants}} \times 100\%$

$= \dfrac{M(C_6H_5COOH)}{M(C_6H_5COOC_2H_5) + M(H_2O)} \times 100 = \dfrac{122.06}{(150.1 + 18.02)} \times 100 = 72.6\%$

Grade boost

Atom economy is obtained from the chemical equation for the reaction.

Atom economy = $\dfrac{\text{mass of required product}}{\text{total mass of reactants}} \times 100\%$

Grade boost

% yield is calculated from the mass of product actually obtained by experiment.

% yield = $\dfrac{\text{mass (or moles) of product obtained}}{\text{maximum theoretic mass (or moles)}} \times 100\%$

quickfire

⑩ Calculate the atom economy for the production of nitric oxide in the oxidation of ammonia.

$$4NH_3 + 5O_2 \longrightarrow 4NO + 6H_2O$$

Percentage error

Usually in AS Chemistry the error is estimated from the uncertainties in the equipment used, such as burette, pipette, balance and thermometer.

Typically the error is taken as one-half of the smallest division on the apparatus, such as 0.05 cm^3 on a burette, $0.1°$ on a $0.2°$ thermometer and 0.0005 g (0.5 mg) on a three-place balance. Remember that with most equipment, since the difference between the initial and final readings is required, two readings are taken so the errors will be 0.1 cm^3 for a burette, $0.2°$ for a $0.2°$ thermometer and 0.001 g for a three-place balance. The error for a standard 25.0 cm^3 pipette is 0.06 cm^3.

These errors are now expressed as percentage errors by dividing by the quantity being measured, giving for example

burette – volume used 24.20 cm^3 (error 0.1 cm^3)

$$\% \text{ error} = \frac{0.10}{24.20} \times 100 = 0.413\%$$

thermometer – ΔT $7.0°$ (error $0.2°$)

$$\% \text{ error} = \frac{0.20}{7.0} \times 100 = 2.86\%$$

balance – 3.610 g (error 0.001 g)

$$\% \text{ error} = \frac{0.001}{3.610} \times 100 = 0.0277\%$$

pipette – volume used 25.0 cm^3

$$\% \text{ error} = \frac{0.06}{25.0} \times 100 = 0.240\%$$

Example

Frances performed a titration experiment to find the concentration of aqueous sodium hydroxide. She pipetted 25.0 cm^3 of aqueous sodium hydroxide into a conical flask and this required 28.65 cm^3 of 0.1250 mol dm^{-3} hydrochloric acid for complete neutralisation. She calculated the concentration to be 0.14325 mol dm^{-3}.

Calculate the percentage error caused by the equipment and state the concentration to a sensible number of significant figures.

$$\text{Pipette \% error} = \frac{0.06}{25.0} \times 100 = 0.24\%$$

$$\text{Burette \% error} = \frac{0.10}{28.65} \times 100 = 0.35\%$$

Total error = 0.59 %

Therefore concentration lies between 0.1424 and 0.1441 mol dm^{-3} and a sensible answer is 0.143 mol dm^{-3}, i.e. three significant figures.

Grade boost

If an error is about three times larger than any other errors there is no need to consider the others at AS level.

Grade boost

If an error is between 0.1% and 1% then a sensible number of significant figures to use is 3.

Significant figures, rounding up and truncation

Most errors in coursework will be around 1% leading to use of three significant figures, but there is no harm in putting a fourth digit in.

Full calculator values may be used at intermediate stages in a calculation; however, it is very important that the full output is not recorded as a final answer. When the calculator answer gives more than the number of significant figures – say three – the third figure may be rounded up only if the fourth figure is 5 or above. Thus 4.268 may be rounded up to 4.27 which is more accurate than 4.26.

A more serious error is to destroy information gained in the experiment by over-shortening the result, e.g. regarding a concentration of 0.0946 mol dm^{-3} as 0.09 or even 0.1!

These are the rules for working out significant figures:

- Zeros to left of first non-zero digit are not significant, e.g. 0.0003 has one sig. fig.
- Zeros between digits are significant, e.g. 3007 has four sig. figs.
- Zeros to the right of a decimal point with a number in front are significant, e.g. 3.0050 has five sig. figs.

Always use significant figures and not decimal places when considering errors. A value of 0.044 has three decimal places but only two significant figures. Stating the value to two decimal places would give a different value (0.04) than the correct significant figures (0.044).

Both standard form and ordinary form are commonly used in calculations. Standard form is particularly useful when large and small numbers are encountered. For example, it is much easier to write 1.25×10^7 than 12 500 000 or 6.2×10^{-6} than 0.0000062. However, it is important that the same number of significant figures is used in both, e.g. 2.63×10^{-3} in standard form, 0.00263 in ordinary form, both have three significant figures.

quickpire

⑪ State the number of significant figures for the following numbers:
a) 0.00625
b) 5010
c) 0.06250

1.4 Bonding

Key Terms

Covalent bond has a pair of electrons of opposed spin shared between two atoms.

Co-ordinate bond is a covalent bond where both electrons come from one of the atoms.

Ionic bond is formed by the attractions between positive ions (cations) and negative ions (anions).

Chemical bonding

Only the inert gases exist as single atoms on the Earth, the atoms of all the other elements being bonded together by electrical forces to form molecules with energy being released in the process. The bonding is of three main types, **ionic, covalent and metallic.**

Bonds may be represented by 'dot and cross' diagrams in which the outer electrons of one atom forming the bond are shown as dots or open circles and those of the other atom by crosses.

Covalent bonding – each atom gives one electron to the (single) bond pair having opposed spins.

Co-ordinate bonding – a covalent bond in which both electrons come from the same atom as in NH_4^+

Ionic bonding – one atom gives one or more electrons to the other and the resulting cation and anion attract one another electrically. Usually there are many ions together in a solid lattice.

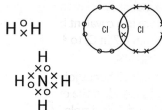

Energy is needed to form the ions from atoms but this is more than repaid by the strong electrical attraction between cation and anion.

Metallic bonding appears under Solids on p41.

▲ Grade boost

Be very careful when drawing 'dot and cross' diagrams to include all electrons, distinguish between electrons from the different atoms and include any charges. You may be asked to omit any inner electrons.

quickfire

① Draw one simple example of each of the three types of bond.

Attractive and repulsive forces

All bonding results from electrical attractions and repulsions between the protons and electrons, with attractions outweighing repulsions.

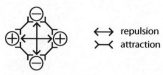

↔ repulsion
⤜ attraction

In covalent bonds the electrons in the pair between the atoms repel one another but this is overcome by their attractions to BOTH nuclei. If atoms get too close together, the nuclei and their inner electrons will repel those of the other atom so that the bond has a certain length. In addition, the electron spins must be opposite for the bonds to form.

In ionic bonding, cations and anions are arranged so that each cation is surrounded by several anions and vice versa to maximise attraction and minimise repulsion.

Electronegativity and bond polarity

In a covalent bond the electron pair is not usually shared equally between the two atoms unless they are the same. Thus one atom will take up a slightly negative charge, the other becoming slightly positive and the bond is now said to be **polar**. These small charges are written over the atoms using the symbols $\delta+$ and $\delta-$ as shown below:

$$\overset{\delta+}{H} \text{------} \overset{\delta-}{F}$$

Co-ordinate bonds are always polar, since the atom giving both electrons to the bond cannot completely lose its rights over one electron.

Bond polarity is governed by the difference in **electronegativity** between the two atoms forming the bond. Electronegativity is a measure of the ability of an atom in a covalent bond to attract the electron pair and, on one scale, ranges from 0.7 in Cs to 4.0 in F, with Cs being electropositive and F highly electronegative.

Thus almost all bonds joining atoms that are not identical will be polar to some extent and, at the other extreme, most ionic bonds have some covalent character, but it is simpler for us to treat ionic bonds as completely ionic and take covalent bonds as having varying degrees of polarity.

The following diagrams show the electron density distribution for some different bonds.

Bond	NaF	H–Cl	H–H
Electronegativity difference	3.1	0.9	0.0
Electron distribution			

The bond in hydrogen chloride, for example, is about 19% ionic.

Key Term

Electronegativity is a measure of the ability of an atom in a covalent bond to attract the electron bonding pair.

Grade boost

As you will see from the Quickfire question, it is the difference between the electronegativity values that is important and not their actual values.

Pointer

Note that partial charges are shown by using the Greek lower case delta δ.

quicKfire

② Using the electronegativity values given, arrange the bonds below in order of INCREASING polarity.
Values: H 2.1, Be 1.5
O 3.5, I 2.5, Cl 3.0
Bonds: I-Cl, H-Cl, Cl-Cl, Be-O

Forces between molecules

Key Terms

Dipole is separation of charge within a molecule. Electrical charges are not balanced so that one part has a partial negative charge and another an equal positive charge.

van der Waals forces are weak intermolecular forces made up of dipole–dipole and induced dipole–induced dipole forces of attraction.

Grade boost

Careful and accurate diagrams are a good way of getting marks in this section and the next one on hydrogen bonding.

quickfire

③ Explain why liquid nitrogen boils at 77K, although the bond between nitrogen atoms in the molecule is very strong.

It is most important to distinguish bonding **BETWEEN** molecules – **INTERMOLECULAR** bonding – and bonding **WITHIN** molecules – **INTRAMOLECULAR** bonding.

Intermolecular bonding is weak and governs physical properties such as boiling temperature; bonding within molecules is strong and governs chemical reactivity. In methane, for example, the forces between the molecules are very weak and the molecules separate, i.e. the liquid boils at −162 degrees, but the C–H bonds are very strong and need a temperature of around 600 degrees before they will break.

Intermolecular bonding is caused by electrical attraction between opposite charges. Although the molecule may be neutral overall, it contains positive and negative charges (electrons and protons) and if the electronegativities of the atoms in the molecule are not the same, the molecule will have a **dipole** with parts that are relatively positive and negative in charge. If these dipoles arrange themselves so that the negative region of one molecule is close to the positive region of another molecule, there will be a net attraction between them.

$$\delta+ \textemdash\, \delta- \qquad \delta+ \textemdash\, \delta-$$
$$\delta- \textemdash\, \delta+ \qquad \delta- \textemdash\, \delta+$$

δ is a permanent partial charge

Even molecules with no dipole show intermolecular bonding, e.g. helium atoms come together to form a liquid at 4K. This is because the electrons are in constant motion around the nuclei so that the centres of positive and negative charge do not always coincide and give a fluctuating dipole. These come into step with one another as one dipole induces an opposite dipole in a nearby molecule giving an attraction between them.

$$\delta\delta+ \textemdash\, \delta\delta- \qquad \delta\delta+ \textemdash\, \delta\delta-$$
$$\delta\delta- \textemdash\, \delta\delta+ \qquad \delta\delta- \textemdash\, \delta\delta+$$

$\delta\delta$ is a fluctuating induced charge

To sum up we have here two types of intermolecular bonding, the first dipole–dipole and the second induced dipole–induced dipole and these two together are called **van der Waals forces**.

Strength

Bonding inside molecules is some 100 times stronger than between them with van der Waals strengths being around 3kJ per mol.

Hydrogen bonding

This is a special intermolecular bonding force that only occurs between molecules that contain hydrogen atoms bonded to very electronegative elements having lone pairs, such as fluorine, oxygen and nitrogen. Although weak compared to bonding inside molecules, it is much stronger than van der Waals forces. Typical strengths for hydrogen bonds would be 30 kJ per mol as against 3kJ for van der Waals and 300kJ for bonding within molecules.

Hydrogen bonding is stronger than that of van der Waals since the small hydrogen atom is sandwiched between two electronegative elements and allows close approach.

We see that the hydrogen atom is especially δ+ being attached to the electronegative oxygen atom so that the oxygen atom on the other molecule is attracted closely to it. Also the bonding is strongest when the three atoms are in a straight line, and note that the internal O–H bond in the molecule is shorter than the dotted hydrogen bond connecting to the other molecule. Since oxygen has two lone pairs and two hydrogen atoms, a tetrahedral hydrogen-bonded structure is formed.

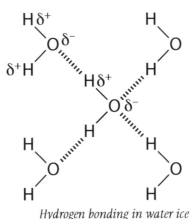

Hydrogen bonding in water ice

Effects of hydrogen bonding on boiling temperature and solubility

Melting and, especially, boiling temperatures increase with the strength of intermolecular forces. With van der Waals forces there is a steady increase with molecular mass and also as dipoles become larger.

The boiling temperature diagram given shows that the stronger hydrogen bonding completely bucks the trend.

In water the molecules freely hydrogen bond with neighbours, these bonds must be largely broken before boiling can occur and thus more energy, i.e. a higher temperature, is needed. There are many similar examples.

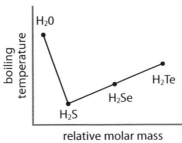

Key Term

Hydrogen bond is a relatively strong intermolecular bond having a hydrogen atom joined to a very electronegative element in a molecule and bonding to another electronegative element in another molecule.

Grade boost

1. Draw a very careful diagram of, e.g., O–H---N, showing the delta plus and delta minuses and the longer dotted H bond.
2. Make it very clear that a hydrogen bond is only a RELATIVELY strong bond compared with a van der Waals bond.

quickfire

④ Arrange the following types of bonding in order of increasing strength: covalent bond inside a molecule, van der Waals bond, hydrogen bond.

quickfire

⑤ Draw a possible hydrogen-bonded structure for liquid ammonia.

Shapes of molecules

The shapes of covalent molecules with more than two atoms and their ions are governed by the electron pairs around the central atom such as C in CH_4. These may be **bonding pairs** holding the atoms together in covalent bonds or **lone pairs** on the central atom that are not usually involved in covalent bonding.

Key Terms

Bond pair is two electrons having opposed spins that bond two atoms in a molecule together by a covalent or co-ordinate bond.

Lone pair is two electrons having opposed spins that belong to one atom only and are not involved in bonding to another atom.

$$\begin{array}{c} H \\ o\!\times \\ H\ o^{+}\ B\ ^{+}o\ H \\ 120° \end{array} \qquad \begin{array}{c} o\ o \\ o\ O\ o \\ H\ ^{+}_{o}\ ^{+}_{o}\ H \\ 104.5° \end{array}$$

Since all electron pairs repel one another, the molecular shape taken up is that allowing the pairs to keep as far away from each other as possible to minimise the repulsion energy. While bonding pairs are spread out between the two atoms bonded in the molecule, lone pairs stay close to the central atom and so repel more than bonding pairs, giving the repulsion sequence:

lone pair – lone pair > lone pair – bond pair > bond pair – bond pair.

Thus in NH_3 with one lone pair and three bonding pairs the repulsion between the lone pair and the bond pairs is greater than that between the bond pairs themselves so that the H–N–H angle is closed up from 109° to 107°.

quickfire

(6) Insert and label the bond pairs and lone pairs in the molecule below:

$$\begin{array}{c} o\ o \\ o\ N\ ^{o}_{\times}\ ^{o}\ H \\ H\ ^{o}_{\times}\ \ o\ H \\ H \\ 107° \end{array}$$

These ideas are applied in the Valence Shell Electron Pair Repulsion (VSEPR) Theory that follows.

VSEPR theory

Valence Shell Electron Pair Repulsion (VSEPR) theory lets us predict the shape of simple molecules in which bonded atoms are arranged around a central atom. The valence shell is the electron shell in which bonding occurs. In VSEPR the number of electron pairs is first found to give the general shape of the molecule, since the repelling pairs keep as far away from one another in space as possible. It is worth repeating the repulsion sequence on the shapes of molecules from page 38:

lone pair – lone pair > lone pair – bond pair > bond pair – bond pair

and seeing how it applies in the figure on page 38.

The shape follows directly from the number of pairs as below:

No of pairs	Shape	Bond angle	Example
2	linear	180°	$BeCl_2$
3	trigonal planar	120°	BF_3
4	tetrahedral	109.5°	CH_4
5	trigonal bipyramid	90°/120°	PCl_5
6	octahedral	90°	SF_6

Secondly, the exact angles between the bonds will change somewhat depending on the repulsion sequence above. Thus in water with two lone pairs the normal tetrahedral bond angle of around 109° for H–O–H is repelled down to 104° by lone pair – bond pair repulsion.

You should be able to predict the shape of any simple molecule given its formula using VSEPR. Note that the same rules apply where the covalent molecule is an ion such as NH_4^+ where all electrons are now in bond pairs and the H–N–H bond angles are all 109.5°.

Note also that you are asked:

- To know the bond angles associated with linear, trigonal, planar, tetrahedral and octahedral molecules and ions.
- To **know** and explain the shapes of BF_3, CH_4, NH_4^+ and SF_6.
- To predict and explain the shapes of other simple species having up to six electron pairs in the valence shell of the central atom.

 Pointer

Learn and understand the shapes for each number of bonding pairs.

 Grade boost

Count the electron pairs carefully and remember that it is only the pairs around the central atom that affect the shape, so that in CF_4, for example, all the lone pairs on the fluorine atoms are ignored, leaving just the four bonding pairs and a tetrahedral shape.

quicKfire

⑦ Predict the general shapes of BH_4, CH_3I, PCl_5 and BeI_2.

Solubility of compounds in water

Key Terms

Solute is the substance that dissolves in the solvent (may be solid, liquid or gas).

Solvent is the liquid medium in which the solute dissolves, commonly water.

Saturated is a solution that cannot dissolve any more solute under the existing conditions.

Polar is a molecule with some separation of positive ($\delta+$) and negative ($\delta-$) charge, e.g. $H^{\delta+}$-----$Cl^{\delta-}$.

> **Pointer**
>
> A useful phrase to remember is 'like dissolves like'.

Grade boost

Clear diagrams with accurate δs and ion charges are good.

NB The Specification here refers to solubility in connection with hydrogen bonding but it is an important general qualitative and quantitative concept and is thus dealt with more fully.

A substance B will be soluble in C if the attractions between B and C molecules are greater than those between B and B and between C and C molecules.

If B is a hydrocarbon and C is water this will not be true and 'oil and water do not mix' but form separate layers. The water molecules attract one another through hydrogen bonding (see above) and hydrocarbon molecules interact more weakly through van der Waals forces. However, in alcohols O–H bonds have been added to the hydrocarbon structure so that hydrogen bonding now occurs between the alcohol and water and the alcohol dissolves in the water provided that its hydrocarbon chain is not too long.

Thus ethanol and the smaller alcohols dissolve in water up to a carbon chain length of about four atoms. Water dissolves many inorganic salts such as NaCl due to a strong interaction between the ions in the salt and the water dipoles.

The **polar** water has $\delta-$ oxygen atoms and $\delta+$ hydrogens, the oxygen regions of several water molecules align themselves around cations and the hydrogens around the anions. The attractions of all these are sufficient to pull the ions from the solid lattice so that the salt dissolves in water.

Solubility terms

Solubility may be expressed as concentrations in either grams/dm³ or mols/dm³ and solubility in mol dm⁻³ = solubility in g dm⁻³ divided by the molar mass.

A **saturated solution** is one that has the maximum possible concentration of **solute** at the given conditions. Knowing the actual concentration is an important part of all chemical procedures.

Soluble salts may be recovered from aqueous solution by, e.g., evaporating water from a warm solution until it is saturated and then cooling to allow the salt to crystallise out.

1.5 Solid structures

Ionic, covalent and metallic

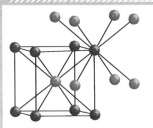

NB. Anions are larger than cations so only the cation size controls the crystal co-ordination number.

CsCl structure   *NaCl*

You should *know* and be able to describe the ionic crystal structures of NaCl and CsCl, the covalent structures of diamond and graphite and know that iodine forms a molecular crystal and know its general structure and that of ice.

In ionic halides, oppositely charged ions pack around one another in such a way as to increase the bonding energy by maximising electrostatic attraction and minimising repulsion. Thus each cation is surrounded by 6 or 8 anions and vice versa, the actual number depending only on the relative sizes in the chlorides. Eight chloride anions can fit around the larger Cs^+ ion while the smaller Na^+ can only accommodate six. The **crystal co-ordination numbers** are therefore 8:8 in CsCl and 6:6 in NaCl.

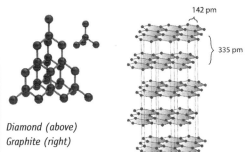
142 pm
335 pm

With diamond, the tetrahedral strong covalent arrangement and build up of a three-dimensional giant structure should be shown while, with graphite, layers made up of covalent hexagons held together by weak forces are needed.

Diamond (above)
Graphite (right)

In solid iodine it is very important to distinguish between the strong covalent bonds holding iodine atoms together in the I_2 molecule and the weak intermolecular forces that hold the I_2 units in the molecular crystal.

Metals

While the actual structures of metals are not required, you need to understand the general concept that atoms of metallic elements each donate one or more electrons to form a delocalised electron sea or gas that surrounds the packed positive ions so formed and binds them together through the attraction between opposite charges.

Key Term

Crystal co-ordination number is the number of anions around each cation in an ionic lattice and vice versa.

Grade boost

While artistic skill is not important in drawing the structures, the charges on all ions must be shown and the crystal co-ordination number must be evident in the diagrams.

Grade boost

A common source of error is not to make it clear that there are two types of bonding in solid iodine, the strongish covalent bond holding the I atoms in I_2 and the weak van der Waals forces holding the I_2 units in the crystal.

Grade boost

You are not expected to have any detailed knowledge of the structures of graphene or carbon nanotubes but up-to-date examples of new properties and uses found in the literature or online will help to impress the examiners. Similarly, good examples of smart materials could be valuable.

Pointer

Build on a solid knowledge of the structures of NaCl, CsCl, diamond and graphite, the fact that two bond types govern the properties of iodine and a qualitative concept of metallic bonding. General physical properties will then be governed by the weakest bonds in the solids and electrical conduction by the need for any charges – ions or electrons – to be mobile. Only in metals are electrons not bound to nuclei and ions are only mobile in the melt, solution or gas.

Grade boost

Get clear in your mind how the properties of solids relate to their molecular structures, and have some examples ready.

New carbon, graphene, nanotubes and buckyballs

Although not in the Specification new forms of carbon are attracting great interest and an awareness of them may have Grade Boost value. Graphene is a graphitic monolayer having outstanding properties and potential, nanotubes are cylindrical rolls of graphene and buckyballs (Buckminster fullerenes) are soccer ball C_{60} structures.

Structure and physical properties

It is important to be able to explain the properties of all the solid types above in terms of their structures.

The giant ionic structures such as the chlorides are generally hard, brittle and high-melting due to the strong ionic bonds. There is no electrical conduction in the solid state since the ions are fixed in the crystal but the molten salts and aqueous solutions of them do conduct since they are now free to move when a voltage is applied. Ionic solids may or may not be soluble in water, depending on energetic or chemical reaction factors, but most ionic chlorides are soluble.

The covalent giants, diamond and graphite, are very high-melting and insoluble in water; diamond is very hard, with each carbon atom being covalently bonded to four others and forming a three-dimensional structure in space, but the weak layer structure in graphite renders it softer and useful as a lubricant. Also graphite conducts electricity owing to the π electron delocalisation in the ring plane while diamond and iodine do not. Iodine is soft and volatile since the I_2 units are held together only by weak van der Waals forces.

In carbon nanotubes the rolled graphene layer structure makes them the strongest and stiffest materials known. As with graphite, their properties are not usually the same in different directions in space owing to the nature of the bonding.

Electron delocalisation in metals gives good electrical and thermal conductivity but their melting temperatures and hardness increase with the number of electrons per atom involved in bonding, e.g.

Metal	Na	Ca	V
no of bonding electrons	1	2	5
melting temp. °C	98	850	1900

Sodium is a soft metal that is easily cut with a knife but vanadium is very hard. Also remember mercury, melting temperature minus 39°C!

In practice the actual properties of a solid depend not only on the bonding at the atomic level but on the way in which the units are held together.

1.6 The Periodic Table

Basic structure

An understanding of the general trends in properties and behaviour gives us great predictive power. The chemistry of the elements is governed largely by their outer electrons so that arranging elements in groups according to their outer structure simplifies study of their behaviour.

Ionisation energy (IE) and electronegativity (x) increase diagonally across the Table (i.e. across a period or up a group), e.g. IEs for Cs and F are 376 and 1680 kJ.

Electrons are thus readily lost in the s block giving cations in ionic compounds; entering the p block in Group 3, IEs become too high so that electron sharing (covalency) is usual, but the more electronegative elements of Groups 6 and 7 can accept electrons to form anions in ionic compounds.

Valency normally rises with group number to a maximum of four and then falls (8 minus the group number) to one in Group 7.

Elements are generally metals when IEs are low in the left and lower regions of the Table and the d block, transition, elements and non-metals in the high IE, upper right portion: semiconductor elements, e.g. Si, are found between these two regions.

Melting temperature trends are more complex, depending on atomic mass, type of solid structure and bond type but decrease down Group 1, rise down Group 7, increase across a period up to Group 4 (C > 3500°C) and then drop sharply as elements form diatomic molecules held by weak intermolecular forces.

	s block		d block	p block				
Group	1	2	transition metals	3	4	5	6	7	8 inert
Oxidn no	1	2					−2	−1	gas
Redox	reducing			oxidising					
Ions	cations			anions					
Oxides	basic			acidic					
MT	decrease down		increase to Group 4	inc. down					
Element type	metals			non-metals					

Trends in the Periodic Table

» Pointer

See also sections on ionisation energies, electronic structures and orbital shapes.

quickfire

① Two of the ions below are not met in ordinary chemistry. Identify these two and state why they are not seen. Ca^{2+} B^{3+} C^{4+}

quickfire

② Explain why IEs increase across a period but decrease down a group.

Key Terms

A **reducing agent** gives an electron to another species and is therefore oxidised by its loss.

An **oxidising agent** removes an electron from another species and is therefore reduced.

Redox is a chemical reaction in which an electron is transferred from one species – the reducing agent – to another species, which is reduced by receiving the electron.

Grade boost

Note that the words redox and oxidation have not necessarily anything to do with oxygen but only with electron transfer.

quickfire

③ a) Write the oxidation numbers for the four species in the reaction $Na + \frac{1}{2}Cl_2 = Na^+ + Cl^-$.

b) Which one of the following correctly gives the oxidation state of the chloride ion?
1-, -I, 0 I.

c) Assign oxidation numbers to the elements in the following compounds: Na_2SO_4, NF_5, O_3, C_2H_6.

Redox

Many chemical reactions involve the loss or gain of electrons, a species being **oxidised** if it loses electrons and **reduced** if it gains them. Since electrons do not vanish or appear from nowhere, all these reactions involve a **transfer** of electrons from the species being oxidised to the one being reduced, e.g. in

$$Na + \frac{1}{2}Cl_2 = Na^+ + Cl^-$$

the Na is oxidised losing an electron and the Cl is reduced gaining an electron.

The popular mnemonic OILRIG is helpful if used carefully as with the atom speaking 'oxidised I lose electrons, reduced I gain electrons'. Confusion is very common.

Oxidation numbers (states)

This is a useful accounting system for **redox** with simple rules:

1 All elements have an oxidation number of zero.

2 Hydrogen in compounds is usually +1 or I.

3 Oxygen is usually −2 or −II.

4 Group 1 and 2 elements in compounds are I and II respectively.

5 Group 6 and 7 elements in compounds are usually −II and −I respectively.

6 An element bonded to itself is still 0.

7 The oxidation numbers of the elements in a compound or ion must add up to zero or the charge on the ion.

IMPORTANT

The oxidation number does not imply a charge, e.g. in MnO_4^- the oxidation numbers are Mn (+7) O_4 (−2 × 4) giving an overall charge of minus 1. The Mn is **not** 7+.

Watch out for the common error described in the Grade Boost.

Evaluate the oxidation numbers of the stated atoms in the following compounds:

H in H_2,

Cr in K_2CrO_4 and in $K_2Cr_2O_7$,

S in $Na_2S_2O_3$ and in $Na_2S_4O_6$,

Both S atoms in the thiosulfate ion of structure $S\text{-}SO_3{}^{2-}$.

Trends in properties of s-block elements Groups 1 and 2

s-block elements

The elements are all reactive electropositive (low electronegativity) metals forming cations with oxidation numbers 1 or 2 respectively.

Oxides are formed with oxygen/air as in $Ca + \frac{1}{2}O_2 \rightarrow CaO$.

Hydrogen is liberated with water and an oxide or hydroxide formed;
$Na + H_2O \rightarrow NaOH + \frac{1}{2}H_2$.

The reaction of Group 2 elements with acids is similar except that a salt is formed as in $Mg + 2HCl \rightarrow MgCl_2 + H_2$ and the elements in both groups show their typical reaction as reducing agents, donating electron(s) to reduce the acid or water to hydrogen and being themselves oxidised.

$$Mg \quad + \quad 2HCl \quad = \quad Mg^{2+} \quad + \quad H_2 \quad + \quad 2Cl^-$$
Oxidn no. $\quad$ 0 $\qquad$ 2(1)(−1) $\qquad$ 2 $\qquad$ 0 $\qquad$ 2(−1)

In all these cases reactivity increases down the group and Group 1 elements are more reactive than Group 2. Lithium reacts slowly with water while potassium is violent; magnesium reacts slowly while barium is faster. All s-block metals react vigorously with acids, react with oxygen and burn in air, while caesium inflames spontaneously.

The oxides and hydroxides are all basic, i.e. they react with acids to give salts, as in $CaO + 2HCl \rightarrow CaCl_2 + H_2O$.

Remember that the Group 2 hydroxide formulae are $M(OH)_2$ since (OH) is −1, i.e. $[O(-2)H(+1)]^{-1}$

While Group 1 salts are all soluble the reactions of Group 2 ions with OH^-, CO_3^{2-} and SO_4^{2-} give a variety of results that must be known. $Mg(OH)_2$ is insoluble in water but solubility increases *down* the group; $BaSO_4$ is insoluble and solubility increases *up* the group; these changes are by factors of hundreds. All the Group 2 carbonates are insoluble and the nitrates soluble.

Flame colours: All of the common elements of Groups 1 and 2 except Mg show characteristic flame colours that must be *known* and that are useful in qualitative analysis.

You should be aware of the great importance of calcium carbonate in both living and inorganic systems and of calcium phosphate minerals in living bones and skeletons. Calcium and magnesium ions play a vital role in the biochemistry of living systems – chlorophyll, muscle operation, etc., and the carbonates exist in huge amount in rocks – chalk, limestone, dolomite.

>> Pointer

A thorough understanding of the concepts and trends in 1.6 will make the work here much easier.

Grade boost

Be careful to write correct formulae for Group 1 and 2 compounds, e.g. KOH, $Mg(OH)_2$, Na_2SO_4, $CaCO_3$ – mistakes look bad.

quickfire

④ a) Match the colours given with the correct element.
   Colour: yellow, brick red, apple green, lilac
   Element: Ba, Ca, K, Na

b) For which Group 2 compounds does the solubility in water:
   (i) increase, and
   (ii) decrease down the group?

Key Term

Electronegative element
is one having a strong
affinity for an electron and
thus acting as an oxidising
agent.

The halogens

These reactive, **electronegative elements** typically form anions having an
oxidation state of –I so that oxidation is the usual reaction as in

$$Na(0) + \tfrac{1}{2}Cl_2\,(0) \ = Na^+(+I) \ + Cl^-(-I)$$

with the Na being oxidised and the oxidising chlorine being reduced from
0 to –I.

The tendency to form anions decreases down the group from fluorine to
iodine with fluorine being the most electronegative element.

The melting temperatures of the elements increase down the group from
gaseous fluorine to solid iodine owing to the increasing intermolecular forces
holding the diatomic elements together in a liquid or solid. This increase is
due to the increasing number of electrons in the molecules contributing to
the induced dipole–induced dipole intermolecular force. For the same reason
volatility decreases down the group.

The halogens react with most metals to form halides with the reactivity
decreasing down the group from fluorine to iodine. A similar feature is shown
in displacement reactions in which a halogen higher in the group displaces
one lower in the group from a salt as in

$$Cl_2 + 2NaBr = Br_2 + 2NaCl$$

This essentially reflects the decrease in oxidising power down the group with
chlorine oxidising the bromide ion to bromine and being itself reduced to
chloride

$$Cl_2\,(0) \ + 2Br(-I) \ = 2Cl^-(-I) + Br_2\,(0)$$

The reaction of halide ions with silver ions in dilute nitric acid is important
in qualitative analysis in both organic and inorganic chemistry. The general
reaction is

$$Ag^+(aq) \ + X^-(aq) \ = AgX(s)$$

The precipitate colours are chloride (white), bromide (pale cream) and iodide
(pale yellow) and only the silver chloride dissolves in dilute ammonia. This
gives a simple way of identifying the halogen present.

Grade boost

Be able to explain clearly why
chlorine displaces bromine
from bromides and bromine
displaces iodine from iodides
in terms of the decrease in
oxidising power down the
group.

quickfire

⑤ Write a balanced equation
giving oxidation states for
the action of chlorine on a
solution of KBr.

Uses of chorine and fluorine in water treatment

Water is treated with chlorine gas to make it safe to drink by killing typhoid
and cholera bacteria.

Fluoride is added to toothpaste as sodium fluoride or a fluorosilicate and
sometimes to water to prevent tooth decay and strengthen bones to reduce
osteoporosis.

The following equilibrium is set up with chlorine in water,

$$Cl_2 + H_2O = HOCl + HCl$$

Ethical issues have been raised over such additions to public water supplies
but the effects appear to be largely beneficial at low concentrations.

1.7. Simple equilibrium and acid-base reactions

Reversible reactions and dynamic equilibrium

Not all chemical reactions 'go to completion', i.e. the reactants change completely to form products. Reactions do not only move in the forward direction, some reactions also move in the backward direction and products change back into reactants. These reactions are called **reversible** and are indicated by the ⇌ sign.

Equilibrium is a term used to denote the balance between the forward and reverse reactions. At equilibrium there is no observable change; however, the system is in constant motion. As fast as the reactants are converted into products, the products are converted back into reactants. No changes are apparent on a macro scale (e.g. concentrations of reactants and products are constant) but reactions continue on a molecular scale. This is called **dynamic equilibrium**.

Le Chatelier's principle

An equilibrium only applies as long as the system remains isolated. In an isolated system, no materials are being added or taken away and no external conditions are being altered.

The proportion of products to reactants in an equilibrium mixture is known as the **position of equilibrium**.

The position of equilibrium can be altered by changing:

- Concentration of the reactants or products
- Pressure in reactions involving gases
- Temperature.

A catalyst does not affect the position of equilibrium, but equilibrium is reached faster.

The effect of a change can be predicted using **Le Chatelier's principle**.

Key Terms

Dynamic equilibrium is when the forward and reverse reactions occur at the same rate.

Le Chatelier's principle states that if a system at equilibrium is subjected to a change, the equilibrium tends to shift so as to minimise the effect of the change.

Pointer

A reversible reaction is one that can go in either direction depending on the conditions.

Pointer

The position of equilibrium is the proportion of products to reactants in an equilibrium mixture.

quickfire

① Explain the term *dynamic equilibrium* for a chemical system.

>> **Pointer**

If a question asks what you would observe when an equilibrium is affected by a change in conditions, you will be expected to use the information given and state any colour changes that would occur.

Grade boost

If the concentration of a reactant is increased the position of equilibrium moves to the right and more products are formed.
Increasing the pressure moves the position of equilibrium to whichever side of the equation has fewer gas molecules.
An increase in temperature moves the position of equilibrium in the endothermic direction.

quickfire

② State what you would observe on increasing: a) the pressure, b) the temperature on the following equilibrium system:
$2NO_2(g) \rightleftharpoons N_2O_4(g)$.
brown   colourless
$\Delta H = -24 kJ\ mol^{-1}$
Explain your answer.

Effect of concentration change

Consider the equilibrium:

$$2CrO_4{}^{2-}(aq) + 2H^+(aq) \rightleftharpoons Cr_2O_7{}^{2-}(aq) + H_2O(l)$$

yellow   orange

Adding more acid increases the concentration of H^+ ions, so the system will try and minimise this effect by decreasing the concentration of H^+ ions and the position of equilibrium will move to the right, forming more products and the colour changes from yellow to orange.

If alkali is now added, the concentration of the H^+ ions decreases, the equilibrium moves to the left and the colour changes back to yellow.

Effect of pressure change

Pressure has virtually no effect on the chemistry of solids and liquids. However, it has significant effects on the chemistry of reacting gases.

The pressure of a gas depends on the number of molecules in a given volume of gas. The greater the number of molecules, the greater the number of collisions per unit time, therefore the greater the pressure of the gas.

Consider the equilibrium:

$$N_2(g) + 3H_2(g) \rightleftharpoons 2NH_3(g) \quad \Delta H = -92\ kJ\ mol^{-1}$$

In total, there are 4 moles of gas on the left-hand side and 2 moles of gas on the right-hand side. Therefore the left-hand side is the side at the higher pressure.

If the total pressure is increased, the equilibrium will shift to minimise this increase. The pressure will decrease if the equilibrium system contains fewer gas molecules. Therefore the position of equilibrium moves to the right (4 moles to 2 moles) and increases the yield of ammonia.

Reducing the pressure shifts the position of equilibrium to the left, decreasing the yield of ammonia.

Effect of temperature change

An endothermic reaction absorbs heat from the surroundings, whereas an exothermic reaction gives out heat to the surroundings. For a reversible reaction, if the forward direction is exothermic, the backward direction is endothermic and vice versa.

Again consider the equilibrium:

$$N_2(g) + 3H_2(g) \rightleftharpoons 2NH_3(g) \quad \Delta H = -92\ kJ\ mol^{-1}$$

Since the enthalpy change is negative, the forward reaction is exothermic (and the backward reaction is endothermic). If the temperature is increased, the system will try and minimise this increase. The system opposes the change by taking in heat so the position of equilibrium moves in the endothermic direction. Therefore the equilibrium moves to the left decreasing the yield of NH_3.

In the same way, decreasing the temperature shifts the equilibrium to the right favouring the exothermic direction and increasing the yield of NH_3.

Equilibrium constant

The position of equilibrium in a reversible reaction may be described in precise terms by combining the equilibrium concentrations to give a value for an **equilibrium constant**. It is given the symbol K_c where the subscript c indicates that it is a ratio of concentrations.

In general for an equilibrium: $aA + bB \rightleftharpoons cC + dD$

$$K_c = \frac{[C]^c \, [D]^d}{[A]^a \, [B]^b}$$ where [C] represents the concentration of C at equilibrium, in mol dm^{-3}.

Note that:

- The products are put in the numerator (top line) and the reactants in the denominator (bottom line).

- The concentrations are raised to powers corresponding to the mole ratio in the equation.

- The unit of K_c can vary, it depends on the equilibrium.

For the equilibrium:

$$CO(g) + H_2O(g) \rightleftharpoons CO_2(g) + H_2(g)$$

$$K_c = \frac{[CO][H_2]}{[CO_2][H_2O]}$$

and the units are: $\dfrac{\text{mol dm}^{-3} \times \text{mol dm}^{-3}}{\text{mol dm}^{-3} \times \text{mol dm}^{-3}}$

therefore the units 'cancel out' and K_c has no units.

The equilibrium constant or equilibrium concentration values can be calculated from appropriate data.

Example

For the system:

$$2SO_2(g) + O_2(g) \rightleftharpoons 2SO_3(g)$$

The equilibrium mixture at a certain temperature contained the following concentrations:

$[SO_2] = 2.75 \times 10^{-3}$ mol dm^{-3} $[O_2] = 4.00 \times 10^{-3}$ mol dm^{-3} $[SO_3] = 3.25 \times 10^{-3}$ mol dm^{-3}

Calculate the value of the equilibrium constant, K_c, at this temperature.

$$K_c = \frac{[SO_3]^2}{[SO_2]^2[O_2]} = \frac{(3.25 \times 10^{-3})^2}{(2.75 \times 10^{-3})^2(4.00 \times 10^{-3})}$$

$$K_c = 3.49 \text{ dm}^3 \text{ mol}^{-1}$$

If, for example, the concentration of SO_2 is needed then simply rearrange the expression for K_c.

$$K_c = \frac{[SO_3]^2}{[SO_2]^2[O_2]}$$

$$[SO_2]^2 = \frac{[SO_3]^2}{K_c[O_2]}$$

$$[SO_2] = \sqrt{\frac{[SO_3]^2}{K_c[O_2]}}$$

Grade boost

Remember the value of K_c is constant for a particular equilibrium reaction at a constant temperature. Therefore only a change in temperature can change the value of K_c.

quicKfire

③ For the reaction $N_2(g) + 3H_2(g) \rightleftharpoons 2NH_3(g)$ write an expression for the equilibrium constant, K_c, giving its units.

quicKfire

④ For the equilibrium: $PCl_5(g) \rightleftharpoons PCl_3(g) + Cl_2(g)$ at a certain temperature, the equilibrium concentrations of PCl_3 and Cl_2 were both 0.0162 mol dm^{-3}. The equilibrium constant, K_c, for the reaction at this temperature, has a value of 0.0108 mol dm^{-3}. Calculate the equilibrium concentration of PCl_5.

Key Terms

Key Terms

An **acid** is a proton (H^+) donor.

A **base** is a proton (H^+) acceptor.

A **strong acid** is one that fully dissociates in aqueous solution.

A **weak acid** is one that partially dissociates in aqueous solution.

≫ Pointer

Salts of HCl are chlorides (Cl^-).
Salts of H_2SO_4 are sulfates (SO_4^{2-}).
Salts of HNO_3 are nitrates (NO_3^-).

quickfire

⑤ Write the equation for the reaction between:

a) Magnesium oxide and sulfuric acid.

b) Calcium carbonate and nitric acid.

quickfire

⑥ Differentiate clearly between a weak acid and a dilute acid.

Acids and bases

All **acids** contain the hydrogen ion, H^+. When an acid is added to water, the acid releases H^+ ions into solution, e.g.

$$HCl(g) \longrightarrow H^+(aq) + Cl^-(aq)$$

(In fact the H^+ ion bonds with a water molecule to form the H_3O^+ ion.)

A **base** is a compound that accepts H^+ ions from an acid, e.g.

$$NH_3(aq) + H^+(aq) \longrightarrow NH_4^+(aq)$$

If a base dissolves in water it is called an alkali. When an alkali is added to water it releases OH^- ions into solution, e.g.

$$NaOH(s) \longrightarrow Na^+(aq) + OH^-(aq)$$

Acids react with bases, alkalis and carbonates to form salts, e.g.

$$HCl + NaOH \longrightarrow NaCl + H_2O$$

or $\quad H^+ + OH^- \longrightarrow H_2O$

$$H_2SO_4 + Na_2CO_3 \longrightarrow Na_2SO_4 + H_2O + CO_2$$

Strong and weak acids

Since acids donate H^+ ions in aqueous solution, the more easily an acid can donate H^+ the stronger the acid.

The general equation for dissociation of an acid is given by:

$$HA(aq) \rightleftharpoons H^+(aq) + A^-(aq) \qquad (A^- \text{ represents an anion})$$

For HCl the equilibrium lies far to the right so the equation is written as:

$$HCl(aq) \longrightarrow H^+(aq) + Cl^-(aq)$$

The acid is fully dissociated or ionised and it is described as a **strong acid**.

Many acids are far from fully dissociated in aqueous solution and these are described as **weak acids**, e.g. for ethanoic acid, the equilibrium

$$CH_3CO_2H(aq) \rightleftharpoons CH_3CO_2^-(aq) + H^+(aq)$$

lies to the left. In fact only about four in every thousand ethanoic acid molecules are dissociated into ions.

The words strong and weak only refer to the extent of dissociation and not in any way to concentration. A **concentrated acid** consists of a large quantity of acid and a small quantity of water. A **dilute acid** contains a large quantity of water.

Similarly bases can be classified as strong or weak. An example of a strong base is NaOH and an example of a weak base is NH_3.

The pH scale

The acidity of a solution is a measure of the concentration of aqueous hydrogen ions, H^+. However, these concentrations are very small and vary over a wide range (between 1 and 0.00000000000001 mol dm^{-3}).

The Danish chemist, Sorenson adopted a logarithmic scale to overcome this and called it the pH scale. He defined pH as:

$$pH = -\log_{10}[H^+] \quad \text{where } [H^+] \text{ is the concentration of } H^+ \text{ in mol } dm^{-3}.$$

The negative sign in the equation results in pH decreasing as the aqueous hydrogen ion concentration increases. If the H^+ ion concentration is greater than 10^{-7} mol dm^{-3}, the pH is less than 7.

Using the pH scale the acidity of any solution can be expressed as a simple more manageable number, ranging from 0 to 14. This is much more convenient for the general public when dealing with concepts of acidity.

Universal indicator colour								
	red	orange	yellow	green	green blue	blue	dark blue	purple
pH	0–2	3–4	5–6	7	8	9–10	11–12	13–14
	all these are acids			neutral	all these are alkalis			
	the stronger the acid the lower the pH				the stronger the alkali the higher the pH			

pH values can be calculated from $[H^+]$ values and vice versa.

Examples

Calculate the pH of a solution of 0.020 mol dm^{-3} hydrochloric acid.

Since HCl is a strong acid it is fully dissociated, therefore if the concentration of HCl is 0.020 mol dm^{-3}, then $[H^+]$ = 0.020 mol dm^{-3}.

$$pH = -\log[H^+] = -\log 0.020 = 1.70$$

A solution of hydrochloric acid has a pH of 3.6, calculate the aqueous hydrogen ion concentration in the solution.

$$3.6 = -\log[H^+]$$
$$[H^+] = 10^{-3.6} = 2.5 \times 10^{-4} \text{ mol } dm^{-3}$$

Key Term

$pH = -\log [H^+]$.

Grade boost

The higher the H^+ ion concentration, the lower the pH and the stronger the acid.

quickfire

① Nitric acid is a strong acid. Calculate:

a) The pH of a solution of 0.005 mol dm^{-3} nitric acid

b) The hydrogen ion concentration of a solution of pH 1.3.

Key Term

A **standard solution** is one whose concentration is accurately known.

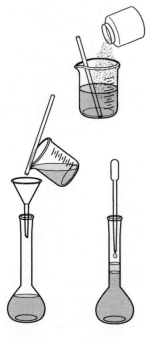

Preparing a standard solution

≫ Pointer

In the preparation of a standard solution:
Use the tare button on the weighing balance so that the scale reads zero.
When adding solid to the weighing bottle, remove the bottle from the pan and then add the solid, checking the mass until the correct amount has been added. This prevents errors caused by spilling solid onto the pan of the balance.

quickfire

⑧ Give two reasons why sodium hydroxide is unsuitable as a primary standard to prepare a standard solution.

Acid-base titrations

An acid-base titration is a type of volumetric analysis where the volume of one solution, say, an acid, that reacts exactly with a known volume of another solution, say, a base, is measured. The precise point of neutralisation is measured using an indicator. One of these solutions must be a **standard solution** (i.e. one of which the exact concentration is known) or it must have been standardised.

In the analysis, you use the standard solution to find out information about the substance dissolved in the other solution.

Preparing a standard solution

A standard solution is prepared using a primary standard. A primary standard is typically a reagent which can be weighed easily, and which is so pure that its weight is truly representative of the number of moles of substance contained. Features of a primary standard include:

1 High purity.

2 Stability (low reactivity).

3 Low hygroscopicity (to minimise weight changes due to humidity).

4 High molar mass (to minimise weighing errors).

A standard solution is prepared from a solid as follows:

- Calculate the mass of the solid required and accurately weigh this amount into a weighing bottle.

- Transfer *all* of the solid into a beaker. Wash out the weighing bottle so that any residue runs into the beaker. Add water and stir until all the solid dissolves.

- Pour all the solution carefully through a funnel into a volumetric (graduated) flask, washing all the solution out of the beaker and the glass rod. Add water until just below the graduation mark.

- Add water drop by drop until the graduation mark is reached and mix the solution thoroughly.

Performing a titration

All titrations follow the same overall method:

- Pour one solution, say, an acid, into a burette, using a funnel, making sure that the jet is filled. Remove the funnel and read the burette.
- Use a pipette to add a measured volume of the other solution, say, a base, into a conical flask.
- Add a few drops of indicator to the solution in the flask.
- Run the acid from the burette to the solution in the conical flask, swirling the flask.
- Stop when the indicator just changes colour (this is the end-point of the titration).
- Read the burette again and subtract to find the volume of acid used (this is known as the titre).
- Repeat the titration, making sure that the acid is added drop by drop near the end-point, until you have at least two readings that are within 0.20 cm³ of each other and calculate a mean titre.

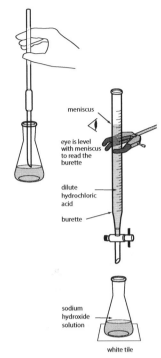

meniscus

eye is level with meniscus to read the burette

dilute hydrochloric acid

burette

sodium hydroxide solution

white tile

Performing a titration

Specified practical work

Practical work is an intrinsic part of chemistry. Although there is no mark as such for laboratory work, the type of direct practical work to be undertaken is listed in the specification and questions will be set on these in the examination. There are four practical exercises in this topic.

Standardisation of an acid solution

This is carried out in two steps. Firstly, prepare a standard solution as described on the previous page, then perform a titration as described above.

Preparation of a soluble salt by titration

A volume of alkali is measured into a flask and a few drops of indicator are added. Acid is added from a burette until the indicator changes colour. When the volume of acid needed to neutralise the alkali has been calculated, the procedure is repeated without the indicator so the correct amount of acid is added to the flask. The solution from the flask is heated to evaporate some of the water. Then it is left to cool and form crystals of pure salt.

Grade boost

Make sure that you can explain why each step is carried out in a titration.

Pointer

In the titration:
When using a pipette, always use a pipette filler – never suck the liquid up by mouth. Never blow the contents of a pipette into the conical flask. The conical flask can be wet as this will not alter the number of moles of alkali/ acid added from the pipette, but the flask must not contain any acid or alkali.
The burette should be rinsed out with a few cm³ of the appropriate acid/alkali.
Record the volume in the burette by looking at where the bottom of the meniscus is on the scale.
Always estimate the volume to the nearest 0.05 cm³.

>> *Pointer*

As part of your practical work,
you should be able to:
• Standardise an acid.
• Perform a back titration.
• Perform a double titration.

>> *Pointer*

For a back titration and
double titration, although the
exact details may differ, the
procedures are the same as for
performing a titration.

Back titration

Sometimes it is not possible to use standard titration methods. For example, the reaction between determined substance and titrant can be too slow, there can be a problem with end-point determination or the base is an insoluble salt. In such situations we often use a technique called back titration.

In back titration a known excess of one reagent **A** reacts with an unknown amount of reagent **B**. At the end of the reaction, the amount of reagent **A** that remains is found by titration. A simple calculation gives the amount of reagent **A** that has been used and the amount of reagent **B** that has reacted.

Double titration

Since different indicators change colour at different pH values, if a solution contains a mixture of two bases which are of different strengths, one titration can be performed, but in two stages, using two different indicators, one added at each stage, to calculate the concentrations of both bases.

For example, the concentrations of sodium hydroxide, NaOH, and sodium carbonate, Na_2CO_3, in a mixture can be determined by titrating with hydrochloric acid, HCl, using phenolphthalein and methyl orange as indicators.

Phenolphthalein changes colour at around pH 9 (pink to colourless).

Methyl orange changes colour at around pH 4 (yellow to orange).

Therefore the first stage of the titration (change in colour of the phenolphthalein) relates to the number of moles of hydroxide and carbonate.

The second stage of the titration (after addition of methyl orange) relates to the number of moles of carbonate only. (Since 1 mol HCO_3^- = 1 mol CO_3^{2-})

Subtracting the moles of carbonate in the second stage from the total moles in the first stage gives the moles of hydroxide. The concentrations of both can then be calculated.

Summary: The Language of Chemistry, Structure of Matter and Simple Reactions

1.1 Formulae and equations

Formulae for compounds and ions

- The group of a metal gives the charge on its ion
- 8 – the group of a non-metal gives the charge on its ion
- For an ionic compound the total number of positive charges must equal the total number of negative charges
- You have to learn the formulae of compound ions and common compounds

Oxidation numbers

When assigning oxidation numbers in a compound remember:

- The sum of the oxidation numbers is zero
- Group 1 metals (and hydrogen) are +1, Group 2 metals are +2
- Oxygen is normally –2

Chemical and ionic equations

- There must be the same number of atoms of each element on each side of a chemical equation
- All formulae must be correct; you cannot change a formula to balance the equation
- In an ionic equation, any ions that do not change during a reaction (spectator ions) are left out. State symbols are normally expected

1.2 Basic ideas about atoms

Radiation

- α particle – cluster of 2 protons and 2 neutrons
- β particle – a fast-moving electron
- γ ray – high energy electromagnetic radiation
- Half-life is the time taken for half of the atoms in a radioactive sample to decay
- Radioactive emissions are potentially harmful since they can damage the DNA of a cell in the body and cause biological damage
- Examples of beneficial uses of radioisotopes are found in medicine, radio-dating and industry

Orbitals or subshells

- An orbital is a region of space in a fixed energy level where there is a high probability of finding an electron
- s subshells contain 1 orbital so can hold 2 electrons
- p subshells contain 3 orbitals so can hold 6 electrons
- d subshells contain 5 orbitals so can hold 10 electrons

Ionisation energy (IE)

- The equation for the 1st IE is: $X(g) \longrightarrow X^+(g) + e^-$
- IE generally increases across a period because of increased nuclear charge
- IE decreases down a group because of increased shielding from inner electrons
- Successive IE measures energy needed to remove each electron in turn until all electrons are removed from an atom

Spectra

- In absorption spectra, energy is absorbed from light causing electrons to move from a lower energy level to a higher one
- In emission spectra, energy is emitted as electrons fall back from a higher energy level to a lower one
- Hydrogen spectrum is a series of discrete lines which get closer as energy increases
- In the hydrogen spectrum each line in the Lyman series (ultraviolet region) is due to electrons returning to the n = 1 energy level, while the Balmer series (visible region) is due to electrons returning to the n = 2 energy level
- Measuring the convergent frequency of the Lyman series (difference from n = 1 to n = ∞) and using $\Delta E = hf$ allows the ionisation energy of the hydrogen atom to be calculated. The value of ΔE is multiplied by Avogadro's constant to give the first molar ionisation energy

1.3 Chemical calculations

Relative mass terms

- Relative atomic mass, A_r, is the average mass of one atom of the element relative to one-twelfth the mass of one atom of carbon-12

  All other relative mass terms are a variation of this

Mass spectrometer

Principles

- Vaporisation of sample before entering mass spectrometer
- Ionisation by bombarding with high energy electrons
- Acceleration – an electric field gets ions to correct speed
- Deflection – a magnetic field separates ions according to their mass/charge ratio
- Detection – ions pass through a slit and are detected by appropriate instruments

Uses

- Determination of relative abundance of isotopes
- Calculating relative atomic masses

Chlorine, Cl_2, spectrum

- Peaks at m/z 35 and 37, due to $^{35}Cl^+$ and $^{37}Cl^+$ respectively, in ratio of 3:1
- Peaks at m/z 70, 72 and 74, due to $(^{35}Cl - ^{35}Cl)^+$ $(^{35}Cl - ^{37}Cl)^+$ and $(^{37}Cl - ^{37}Cl)^+$ respectively, in ratio of 9:6:1

Empirical and molecular formulae

- Empirical formula shows the simplest whole number ratio of the amount of elements present
- Molecular formula is the actual formula of a particular compound; it is a simple multiple of the empirical formula

Moles in calculations

- A mole is the amount of any substance that contains the same number of particles as there are atoms in exactly 12 g of carbon-12
- This is a large number, 6.02×10^{23}, and is called the Avogadro constant, L
- For a solid: number of moles (n) = $\dfrac{\text{mass of substance (m)}}{\text{molar mass (M)}}$
- For a solution: number of moles (n) = concentration (c) × volume (v)
- For a gas: number of moles (n) = $\dfrac{\text{volume of gas (v)}}{\text{molar volume } (v_m)}$ (At 0 °C molar volume = 22.4 dm³)
- To correct volume due to changes in temperature or pressure use: $\dfrac{P_1 V_1}{T_1} = \dfrac{P_2 V_2}{T_2}$ where 1 represents the initial conditions and 2 the final conditions
- Rearranging the ideal gas equation PV = nRT can also be used to calculate the number of moles of a gas

Atom economy and percentage yield

- Atom economy = $\dfrac{\text{mass of required product}}{\text{total mass of reactants}} \times 100$
- Percentage yield = $\dfrac{\text{mass (moles) of product obtained}}{\text{maximum theoretical mass (moles) of product}} \times 100$

Percentage error and significant figures

- Percentage error of any apparatus = $\dfrac{\text{one-half of the smallest division on the apparatus}}{\text{the quantity being measured}} \times 100$
- Normally the difference between the initial and final readings is required therefore this expression is multiplied by 2
- If an error is between 0.1% and 1% then a sensible number of significant figures to use is 3

1.4 Bonding

Bond types

- Ionic bonds formed by the electrical attraction between oppositely charged ions
- Covalent bonds formed by an electron pair with opposed spins
- Coordinate bond is a covalent bond where both electrons come from one atom

Bond polarity and electronegativity

- Most bonds are intermediate between ionic and covalent and are called polar
- Electronegativity measures the electron-attracting power of an atom in a covalent bond; the greater the electronegativity difference between the bond atoms, the more polar is the bond

Intermolecular bonding

- There is weak bonding between all molecules due to permanent and temporary dipoles. These van der Waals forces are much weaker than covalent and ionic bonds but govern physical properties
- Hydrogen bonds are stronger intermolecular bonds formed when hydrogen is bonded between very electronegative F, O and N atoms

Shapes of molecules

- The VSEPR method enables molecular shapes to be predicted from the number of bonding pairs and lone pairs about a central atom. These arrange themselves to minimise repulsions
- The general shapes are linear (2 pairs), trigonal (3 pairs), tetrahedral (4 pairs) and octahedral (6 pairs) with the actual bond angles depending on the numbers of lone pairs and bonding pairs

1.5 Solid structures

Crystal structures

- Giant and molecular crystals: in giant molecules such as ionic and metal crystals and diamond, the whole crystal lattice is really one molecule. In molecular crystals, such as ice, individual (water) molecules are held together in the lattice by weak intermolecular forces.

- Ionic crystals: the structures of sodium and caesium chlorides should be known and the reason for the difference between them understood

- Covalent giant molecules: the structures of diamond and graphite should also be known

- Molecular crystals: the crystal structures of ice and iodine should be understood

- Metals: the 'electron sea' model should be known and understood

- Structure and physical properties: it is important to be able to understand and discuss the physical properties of solids in terms of their structures

- Solubility in water: this important property of solids should also be related to their structures

1.6 The Periodic Table

- Electronic structures and the Table

- Trends in ionisation energy, oxidation states and electronegativity down groups and across periods

- s-Block chemistry; knowledge and understanding of chemistry of Group 1 and Group 2 elements and compounds

- Group 7 chemistry – the halogens

- Halogens uses in water treatment

- Qualitative and quantitative analysis in Practical Work

1.7 Chemical equilibrium and acid-base reactions

Equilibrium

- Dynamic equilibrium is when the forward and reverse reactions in a reversible reaction occur at the same rate

- Position of equilibrium refers to the proportion of products to reactants in an equilibrium mixture

- Le Chatelier's principle states that if a system at equilibrium is subjected to a change, the equilibrium tends to shift in order to minimise the effect of the change

- If the concentration of a reactant is increased, the position of equilibrium moves to the right and more products are formed

- An increase in temperature moves the position of equilibrium in the endothermic direction

- Increasing the pressure moves the position of equilibrium to whichever side of the equation has fewer gas molecules

- An equilibrium constant, K_c, can be calculated for any equilibrium. Only a change in temperature can change the value of K_c

Acids and bases

- Acids are proton donors, bases are proton acceptors

- Strong acids are fully dissociated in solution

- $pH = -\log[H^+]$. It is a measure of acidity using manageable numbers with a scale from 0 to 14

- The lower the pH the higher the H^+ ion concentration and the stronger the acid

- When preparing a standard solution accurate weighing scales and a volumetric flask must be used

- During titrations, the key items of apparatus are burette, pipette and conical flask.

- At least two titres that are within 0.20 cm^3 of each other are required

- When calculating a mean titre only use results that are within 0.20 cm^3 of each other

Energy, Rate and Chemistry of Carbon Compounds

The most important chemical reaction, combustion, is carried out to provide energy; and the energy changes accompanying chemical changes are crucial factors in our understanding of equilibrium. Also of great practical importance is kinetics, the study of the rate at which, and the mechanism by which, reaction takes place.

The social, economic and environmental aspects of chemical processes are considered, in particular fossil and biomass fuels and carbon neutrality, along with the increasingly important role of Green Chemistry.

An introduction to organic chemistry provides a way to understand how the properties of carbon compounds can be modified by the introduction of functional groups. A further topic deals with analytical techniques that use mass spectral data and characteristic infrared frequencies in the elucidation of structure.

Revised it!

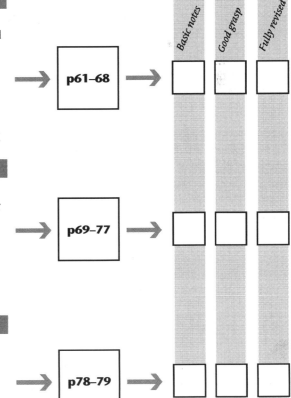

2.1 Thermochemistry

Most chemical reactions are accompanied by enthalpy changes that result in temperature changes. The most useful enthalpy changes are formation, Δ_fH, and combustion Δ_cH. Enthalpy changes that are difficult to measure can be calculated using Hess's law. Enthalpy changes involving covalent compounds can be calculated using average bond enthalpies. The enthalpy change of a reaction can be measured in the laboratory using a simple calorimeter, e.g. a coffee cup calorimeter and using the expression $q = mc\Delta T$.

p61–68

2.2 Rates of reaction

The rate of a chemical reaction can be calculated by measuring the concentration of a reactant (or product) over time. Various methods of measuring reaction rates in the laboratory are shown. Changes to rate during a reaction and changes occurring as a result of varying conditions are explained using collision theory. Energy profiles and energy distribution curves are used to further explain the importance of activation energy and catalysts.

p69–77

2.3 Wider aspects of chemistry

The problems of large-scale chemical manufacture and energy production are considered. Green Chemistry helps to reduce energy use and minimise pollution and waste and the aim of reducing fossil fuel consumption and the associated CO_2 production is examined, especially through kinetics and thermodynamics.

p78–79

2.4 Organic compounds

There are millions of organic compounds and to make sense of their reactions, patterns of behaviour must relate to their structures and the existence of homologous series. Formulae can be shown in different ways according to the use being made of the formula. Many compounds exist as isomers and there are several different types of isomerism. Reactions of compounds within an homologous series are similar but the physical properties alter gradually as the series is descended.

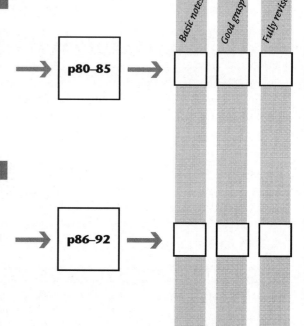

p80–85

2.5 Hydrocarbons

Historically fossil fuels have been used for all our energy needs but they are a finite resource, with disadvantages in their use, so that new energy sources are now being developed. Alkanes and alkenes are both homologous series of hydrocarbons but the difference in electron density in their structures means that they are susceptible to different types of attack. This means that they react by different mechanisms so that alkanes undergo substitution reactions whilst alkenes undergo addition reactions.

p86–92

2.6 Halogenoalkanes

Halogenoalkanes react by yet another type of mechanism – their polarity means they are susceptible to nucleophilic substitution but they can also undergo elimination reactions. CFCs were widely used, due to their properties, but nowadays environmental issues have drastically limited their use and research is going ahead to find replacements.

p93–96

2.7 Alcohols and carboxylic acids

Ethanol is the best-known alcohol and it, like all other alcohols, contains the functional group OH. Most alcohols can be oxidised and dehydrated to produce a variety of organic compounds. Carboxylic acids contain the COOH functional group. Although carboxylic acids are weak, they take part in reactions generally associated with inorganic acids. They also form esters with alcohols.

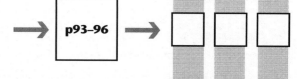

p97–102

2.8 Instrumental analysis

These are the techniques by which the identification of organic compounds is really carried out nowadays. Different spectra give different information about an organic unknown so that generally more than one spectrum is needed to make a definitive identification.

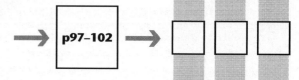

p103–106

Basic notes · Good grasp · Fully revised

2.1. Thermochemistry

Enthalpy changes

There are many forms of energy. We shall be dealing with two: heat, a form of kinetic energy and chemical energy, a form of potential energy.

In a chemical reaction, existing bonds are broken and new bonds are made. This changes the chemical energy of atoms and energy is exchanged between the chemical system and the surroundings, as heat. To measure this energy the term enthalpy, H, is used.

Although enthalpy cannot be measured, an enthalpy change, ΔH, can easily be measured. Its units are joules, J, or kilojoules, kJ.

$$\Delta H = H_{products} - H_{reactants}$$

For exothermic changes (i.e a reaction that releases heat) heat is given out to the surroundings so $H_{products} < H_{reactants}$ and ΔH is negative.

For endothermic changes (i.e a reaction that absorbs heat) heat is taken in from the surroundings so $H_{products} > H_{reactants}$ and ΔH is positive.

All the energy that leaves a system enters the surroundings (and vice versa). In an exothermic reaction, energy is not being created and it is not being destroyed in an endothermic reaction. The total energy of the whole system of reacting chemicals and surroundings is constant. This is an important principle and is called the **principle of conservation of energy**.

Since enthalpy change for reactions depends on the conditions, for values to be compared, standard enthalpy change is measured when fixed conditions are used. The conditions are:

- All substances in their standard states
- A temperature of 298 K (25°C)
- A pressure of 1 atm (101 000 Pa)

The symbol for a standard enthalpy change is ΔH^{θ}.

Three important standard enthalpy changes are dealt with at AS level.

Standard enthalpy change of formation, $\Delta_f H$

This is the enthalpy change when one mole of a substance is formed from its constituent elements in their standard states under standard conditions.

For example, the standard enthalpy change of formation of water is represented as:

$$H_2(g) + \tfrac{1}{2}O_2(g) \longrightarrow H_2O(l) \qquad \Delta_f H = -286 \text{ kJ mol}^{-1}$$

If we are forming an element, such as $H_2(g)$, from the element $H_2(g)$, there is no chemical change. Therefore all elements in their standard state have a standard enthalpy change of formation of 0 kJ mol^{-1}.

Key Terms

Enthalpy, H, is the heat content of a system at constant pressure.

Enthalpy change, ΔH, is the heat added to a system at constant pressure.

>> *Pointer*

The chemical system is the reactants and products. The surroundings are everything other than the system.

Grade boost

Remember standard conditions are 298 K and 1 atm.

Key Term

The standard enthalpy change of formation, $\Delta_f H^{\theta}$, is the enthalpy change when one mole of a substance is formed from its constituent elements in their standard states under standard conditions.

>> *Pointer*

When we write thermochemical equations, we may need to use fractions like '½O₂', so that we have the correct number of moles involved in the change.

Key Term

The standard enthalpy change of combustion, $\Delta_c H^\theta$, is the enthalpy change when one mole of a substance is completely combusted in oxygen under standard conditions.

Standard enthalpy change of combustion, $\Delta_c H$

This is the enthalpy change when one mole of a substance is completely combusted in oxygen under standard conditions.

For example, the **standard enthalpy change of combustion** of methane is given by:

$$CH_4(g) + 2O_2(g) \longrightarrow CO_2(g) + 2H_2O(l) \qquad \Delta_c H = -891 \text{ kJ mol}^{-1}$$

Enthalpy change of reaction, $\Delta_r H$

This is the enthalpy change in a reaction between the number of moles of reactants shown in the equation for the reaction.

There is no reason why a standard enthalpy change of reaction should be related to one mole of reactants or products.

For example, the reaction between ammonia and fluorine is written as:

$$NH_3(g) + 3F_2(g) \longrightarrow 3HF(g) + NF_3(g) \qquad \Delta_r H = -875 \text{ kJ mol}^{-1}$$

The standard enthalpy change for a chemical reaction can be calculated from the standard enthalpy changes of formation of all reactants and products involved. The standard enthalpy of reaction, $\Delta_r H^\theta$ is given by:

$$\Delta_r H = \Sigma \Delta_f H(\text{products}) - \Sigma \Delta_f H(\text{reactants}) \qquad (\Sigma \text{ stands for 'sum of'})$$

For example, for the reaction between ammonia and fluorine:

Compound	$NH_3(g)$	$F_2(g)$	$HF(g)$	$NF_3(g)$
$\Delta_f H^\theta$ / kJ mol^{-1}	−46	0	−269	−114

$$
\begin{aligned}
\Delta_r H &= \Sigma \Delta_f H(\text{products}) - \Sigma \Delta_f H(\text{reactants}) \\
&= (3(-269) + (-114)) - ((-46) + 0) \\
&= -921 + 46 \\
&= -875 \text{ kJ mol}^{-1}
\end{aligned}
$$

(remember that $\Delta_f H$ for any element in its standard state is 0)

Hess's law

It is not always possible to measure the enthalpy change of a reaction directly. **Hess's law**, which is based on the law of conservation of energy, gives a method for finding an enthalpy change indirectly.

The enthalpy cycle shows two routes for converting reactants to products. The first is a direct route and the second an indirect route via the formation of an intermediate.

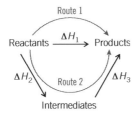

By Hess's law the total enthalpy is independent of the route, so route 1 = route 2

i.e. $\Delta H_1 = \Delta H_2 + \Delta H_3$

If the sum of $\Delta H_2 + \Delta H_3$ were different to ΔH_1, it would be possible to create energy by making the products via the intermediate by one route and then converting back to the reactants by the other route. This would be contrary to the law of conservation of energy.

> **Key Term**
>
> **Hess's law** states that the total enthalpy change for a reaction is independent of the route taken from the reactants to the products.

a. For the reaction: $4NH_3(g) + 5O_2(g) \rightarrow 4NO(g) + 6H_2O(l)$
   the relevant enthalpies of formation are:

Substance	$NH_3(g)$	$O_2(g)$	$NO(g)$	$H_2O(l)$
$\Delta_f H^\ominus$/kJ mol^{-1}	-46	0	90	-286

(i) State why the standard enthalpy change of formation of $O_2(g)$ is zero.

(ii) Calculate the enthalpy change for this reaction.

b. Copper(II) oxide can be reduced to copper by carbon monoxide
   $CuO(s) + CO(g) \rightarrow Cu(s) + CO_2(g)$ $\Delta_r H = -126$ kJ mol^{-1}
   Calculate the enthalpy change of formation of CuO given that
   $\Delta_f H(CO) = -111$ kJ mol^{-1} and $\Delta_f H(CO_2) = -394$ kJ mol^{-1}.

Grade boost

Using enthalpy changes of formation $\Delta H = \Sigma\Delta_f H(\text{products}) - \Sigma\Delta_f H(\text{reactants})$
Using enthalpy changes of combustion $\Delta H = \Sigma\Delta_c H(\text{reactants}) - \Sigma\Delta_c H(\text{products})$

>> Pointer

In an enthalpy cycle using $\Delta_f H$, the direction of the arrows goes from the common elements to the reactants and products.
In an enthalpy cycle using $\Delta_c H$, the direction of the arrows goes from the reactants and products to the common combustion products.

① Given the following enthalpy changes of combustion:

Substance	$\Delta_c H^\theta$/kJ mol^{-1}
$C_3H_7OH(l)$	−2010
C(s)	−394
$H_2(g)$	−286

Calculate the enthalpy change of formation of propan-1-ol

Hess's law enables us to calculate standard enthalpy changes of reactions that would be difficult to measure, by using enthalpy cycles.

For example, the combustion of ammonia:

$$4NH_3(g) + 3O_2(g) \rightarrow 2N_2(g) + 6H_2O(l)$$

Substance	$NH_3(g)$	$O_2(g)$	$NO(g)$	$H_2O(l)$
$\Delta_f H^\theta$/kJ mol^{-1}	−46	0	90	−286

Constructing an enthalpy cycle linking the elements to the reactants and products gives:

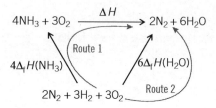

Since $\Delta_f H$ is given, the directions of the arrows go from the common elements to the reactants and products.

By Hess's law, route 1 = route 2

$$\Delta H + 4(-46) = 6(-286)$$

$$\Delta H = -1716 + 184$$

$$\Delta H = -1532 \text{ kJ mol}^{-1}$$

Enthalpy changes of formation of hydrocarbons are impossible to measure practically since burning carbon in hydrogen will produce a mixture of hydrocarbons. However, if enthalpy changes of combustion are known, this problem can be overcome using Hess's law.

E.g. for the reaction $\quad 4C(\text{graphite}) + 5H_2(g) \longrightarrow C_4H_{10}(g)$

Substance	C(s)	$H_2(g)$	$C_4H_{10}(g)$
$\Delta_c H^\theta$/kJ mol^{-1}	−394	−286	−2877

Constructing an enthalpy cycle linking the reactants and products to the common combustion products gives:

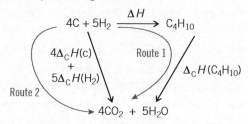

Since $\Delta_c H$ is given, the directions of the arrows go from the reactants and products to the common combustion products.

By Hess's law, route 1 = route 2

$$\Delta H + (-2877) = (4(-394) + 5(-286))$$

$$\Delta H = -3086 + 2877$$

$$\Delta H = -129 \text{ kJ mol}^{-1}$$

Bond enthalpies

Bond enthalpy gives information about the strength of a covalent bond. Bond enthalpies show how much energy is needed to break different covalent bonds.

The H–H bond enthalpy is always the same because the H–H bond only exists in a H_2 molecule. However, C–H bonds exist in many different compounds. The actual value for the enthalpy change for a particular bond depends on the structure of the rest of the molecule, so the C–H bond strength varies across the different environments in which it is formed.

We therefore take average values for bond enthalpy derived from the full range of molecules that contain a particular bond.

Average bond enthalpies can be used to calculate standard enthalpy changes of reaction involving covalent compounds. The results of these calculations will not be as accurate as results derived from experiments with specific molecules. However, they usually give an accurate enough indication of the standard enthalpy change of reaction.

Example

Ethanol is used as a fuel because its combustion reaction is strongly exothermic.

$$C_2H_5OH(l) + 3O_2(g) \rightarrow 2CO_2(g) + 3H_2O(l)$$

Calculate the standard enthalpy change for this reaction using the average bond enthalpies given below.

Bond	C – C	C – H	C – O	C = O	O – H	O = O
Average bond enthalpy/ kJ mol^{-1}	348	413	360	805	463	496

Step 1 Draw out each molecule

Step 2 Calculate the energy required to break the bonds (endothermic).

Bonds broken: $5(C – H) + (C – C) + (C – O) + (O – H) + 3(O = O) =$ 4724 kJ mol^{-1}

Step 3 Calculate the energy released when bonds are made (exothermic).

Bonds formed: $4(C = O) + 6(O – H) = -5998$ kJ mol^{-1}

Step 4 Add together the energy changes

$\Delta H = \Sigma$(bonds broken) $+ \Sigma$(bonds formed)

$\Delta H = 4724 + (-5998) = -1274$ kJ mol^{-1}

Key Terms

Bond enthalpy is the enthalpy required to break a covalent X–Y bond into X atoms and Y atoms, all in the gas phase.

Average bond enthalpy is the average value of the enthalpy required to break a given type of covalent bond in the molecules of a gaseous species.

Grade boost

When asked to calculate enthalpy change using bond enthalpies, always draw out each molecule so that you can see the bonds broken and bonds made.

Pointer

Breaking bonds requires energy, therefore is endothermic (so bond enthalpy is always positive). Making bonds releases energy so is exothermic.

quickfire

② Ethane can be formed from ethene and hydrogen

$$C_2H_4 + H_2 \rightarrow C_2H_6$$

Using average bond enthalpies, calculate the enthalpy change for this reaction:

Bond	Average bond enthalpy/ kJ mol^{-1}
C = C	612
C – H	413
H – H	436
C – C	348

Calculating enthalpy changes

You cannot directly measure the heat content (enthalpy) of a system but you can measure the heat transferred to its surroundings. This process involves carrying out the chemical change in an insulated container called a calorimeter. The change in the temperature inside the calorimeter caused by the enthalpy change of the reaction can be measured with a thermometer.

If the temperature change is recorded and the mass and specific heat capacity of the contents of the calorimeter are known then the enthalpy change can be calculated.

The relationship between the temperature change, ΔT, and the amount of heat transferred, q, is given by the expression:

$$q = mc\,\Delta T$$

m is the mass of the solution in the cup

c is the specific heat capacity of the solution

For the purposes of the calculations, we assume that

- All the heat is exchanged with the solution alone
- The solution has the same specific heat capacity as water ($4.18\ Jg^{-1}K^{-1}$)
- The density of the solution is $1\ g\ cm^{-3}$.

Therefore the mass will be the same as the volume of the solution. The mass of a solid is not added to the mass of the solution. However, if both reactants are solutions then the mass is equal to the total volume of the solutions.

> ## ≫ Pointer
> Specific heat capacity, c, is the energy required to raise the temperature of 1 g of a substance by 1 K. (The value for water is $4.18\ Jg^{-1}K^{-1}$ and will always be given.)

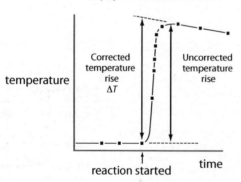

Graph for an exothermic reaction

temperature

Corrected temperature rise ΔT

Uncorrected temperature rise

reaction started   time

To obtain the maximum temperature change, allowances are made for heat lost (or gained) to (or from) the surroundings. Therefore, temperatures of the solution are taken for a short period before mixing and for some time after mixing. A graph of temperature against time is plotted and the maximum temperature is obtained by extrapolating the graph back to the mixing time.

To calculate the enthalpy change of reaction per mole, we use the expression:

$$\Delta H = \frac{-q}{n}$$

Where n is the amount in moles that has reacted.

Since the number of moles is needed to calculate the molar enthalpy change, the reactant that is not in excess has to be measured accurately. So the mass of a solid or the concentration of a solution must be known.

> ## ≫ Pointer
> The minus sign is used in the expression $\Delta H = \frac{-q}{n}$ because if there is an increase in temperature the reaction is exothermic and ΔH is negative.

Example

0.100 g of magnesium were added to 50.0 cm³ of 1.00 mol dm⁻³ sulfuric acid solution in a polystyrene cup. The maximum temperature change (rise) was calculated to be 9.6 °C.

Calculate the enthalpy change for the reaction:

$$Mg(s) + H_2SO_4(aq) \longrightarrow MgSO_4(aq) + H_2(g)$$

Step 1 Calculate the amount of heat transferred in the experiment.

Only 50 cm^3 of solution, therefore mass is 50 g

$q = mc\Delta T = 50 \times 4.18 \times 9.6 = 2006$ J

Step 2 Calculate the amount, in moles, of the reactants

moles Mg $= \dfrac{0.100}{24.3} = 0.00411$

moles $H_2SO_4 = 1 \times 0.050 = 0.050$

therefore Mg is not in excess and is used in the calculation

Step 3 Calculate the molar enthalpy change

$\Delta H = \dfrac{-q}{n} = -\dfrac{2006}{0.00411} = -488\ 078$ J mol^{-1} = -488 kJ mol^{-1}

Specified practical work

There are two practical exercises in this topic.

Determining an enthalpy change of combustion

The experimental determination of $\Delta_c H$ for a fuel can easily be carried out.

A known mass of fuel is burnt in air to heat a known mass of water and the temperature change in the water is recorded.

These are the main points to note in this practical work:

- Allow a suitable gap between the base of the metal container and the top of the spirit burner.
- Accurately measure the amount of water being added to the metal container.
- Use an accurate thermometer to measure the initial temperature of the water. When a steady value has been obtained record the temperature.
- Weigh the spirit burner containing the fuel and record the initial mass.
- After lighting the wick adjust the gap between the metal container and the spirit burner if necessary.
- Allow the fuel to heat the water to a suitable temperature (an increase of about 20 °C is adequate – the smaller the increase, the greater the error on the thermometer).
- Extinguish the flame and record the final maximum temperature.
- Allow the spirit burner to cool thoroughly before re-weighing and recording the final mass.

The value is much lower than the book value because:

- Some of the energy transferred from the burning fuel is 'lost' in heating the apparatus and the surroundings.
- The fuel is not completely combusted.

③ The combustion of 1.50 g of ethanol raised the temperature of 500 cm^3 of water by 19.5°C. Calculate the molar enthalpy change of combustion of ethanol, C_2H_5OH.

quickpire

④ 20.0 cm^3 of 1.000 mol dm^{-3} HCl solution were reacted with 50.0 cm^3 of 0.400 mol dm^{-3} NaOH solution and the maximum temperature rise was calculated as 3.6°C. Calculate the molar enthalpy of neutralisation for this reaction.

Grade boost

Show all your working. Each step of the calculation will be worth a mark.

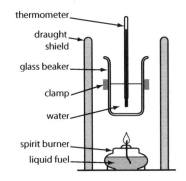

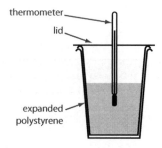

thermometer
lid
expanded polystyrene

>> **Pointer**
Other suitable examples of indirect determination of enthalpy changes are:
Formation of magnesium oxide (adding magnesium and magnesium oxide separately to hydrochloric acid).
Formation of magnesium carbonate (adding magnesium and magnesium carbonate separately to hydrochloric acid).
Decomposition of sodium hydrogencarbonate (adding sodium hydrogencarbonate and sodium carbonate separately to hydrochloric acid).

quickfire

⑤ The enthalpy change of solution of ammonium chloride is 15.2 kJ mol^{-1}. Calculate the temperature change when 7.60 g of ammonium chloride are dissolved in 125 cm^3 of water.

>> **Pointer**
This process is also suitable to find an enthalpy change for displacement reactions and an enthalpy change of solution. However, only one solid would be used and Hess's law would not be used.

Indirect determination of an enthalpy change

The simplest type of calorimeter is a coffee cup calorimeter. It can be used to measure changes that take place in aqueous solution.

The expanded polystyrene insulates the solution inside the cup so the amount of heat lost or absorbed by the cup during the experiment is negligible.

All actual values obtained this way are lower than book values due to heat loss from the simple type of calorimeter used.

An example of an indirect determination of enthalpy change is the enthalpy change of reaction of magnesium oxide with carbon dioxide to form magnesium carbonate. The enthalpy changes of reaction between magnesium and acid and magnesium carbonate and acid are separately measured.

These are the main points to note in this practical work:

- Measure an appropriate volume of acid using a burette or pipette (the mass of acid is used in the expression to calculate ΔH) and place it in a polystyrene cup. This must be in excess (to ensure that all the solid reacts).

- Use an accurate thermometer to measure the initial temperature of the acid. When a steady value has been obtained, record the temperature. (ΔT is used in the expression to calculate ΔH.)

- Accurately weigh the solid, in powder form (to ensure as rapid a reaction as possible), in a suitable container (the amount, in moles, of the solid is used in the expression to calculate ΔH).

- Add all the solid to the cup, stir the mixture well (to ensure that the reaction is as rapid as possible and that all the solid is used) and start a stopwatch.

- Keep stirring with the thermometer and record the temperature regularly (about every 30 seconds). Stop recording the temperature when it has fallen for about 5 minutes.

- Re-weigh the weighing container to ensure that the correct mass of solid added is recorded.

- Plot a graph of temperature against time to calculate the maximum temperature the mixture might have reached. (This is essential to calculate the correct ΔT – see page 66.)

- Calculate the amount of heat transferred (q = mc ΔT).

- Calculate the enthalpy change for the reaction (ΔH = −q/n)

- Repeat the procedure with the other solid.

- Use Hess's law to calculate the required enthalpy change.

2.2. Rates of reaction

Chemical kinetics investigates the rates at which chemical reactions take place.

For a reaction: $\text{rate} = \dfrac{\text{change in concentration}}{\text{time}}$ units: $\dfrac{\text{mol dm}^{-3}}{\text{s}} = \text{mol dm}^{-3}\,\text{s}^{-1}$

If another variable, such as mass or volume is measured, the rate can be expressed in corresponding units such as $g\,s^{-1}$ or $cm^3\,s^{-1}$.

Usually for reactions:

- Rate is fastest at the start of a reaction since each reactant has its greatest concentration.
- Rate slows down as the reaction proceeds since the concentration of the reactants decreases.
- Rate becomes zero when the reaction stops, i.e. when one of the reactants has been used up.

Key Term

The rate of reaction is the change in concentration of a reactant or product per unit time.

Grade boost

When drawing a graph, draw a line that best fits the points. All the points might not be on the 'best fit' line.

Calculating initial rates

We follow the rate of a reaction by measuring the concentration of a reactant (or product) over a period of time. The results obtained are plotted to give a graph. To find the initial rate, it is necessary to find the initial slope (gradient) of the line. At AS level the graph will probably be a straight line to begin with, e.g.

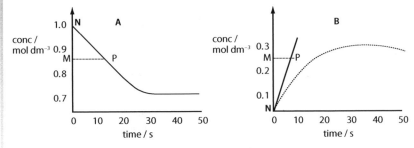

To find the gradient:

For graph A, the line is straight initially, therefore at any convenient point, P, on the line draw a horizontal line MP to the y axis and draw a vertical line from M to the beginning of the slope, N.

$\text{Rate} = \dfrac{\text{change in concentration}}{\text{time}} = \dfrac{MN}{MP} = \dfrac{(1-0.86)}{13} = \dfrac{0.14}{13} = 0.011\ \text{mol dm}^{-3}\,\text{s}^{-1}$

For graph B, the line is curved initially, therefore you have to draw a straight line by drawing a tangent as close as possible to the beginning of the curve.

$\text{Rate} = \dfrac{\text{change in concentration}}{\text{time}} = \dfrac{MN}{MP} = \dfrac{0.25}{7} = 0.036\ \text{mol dm}^{-3}\,\text{s}^{-1}$

quickfire

① For the reaction

$Zn + 2HCl \longrightarrow ZnCl_2 + H_2$ the initial concentration of the acid was $0.300\ \text{mol dm}^{-3}$. After 50 s the concentration had decreased to $0.142\ \text{mol dm}^{-3}$.

(a) Calculate the average rate of reaction over this time interval.

(b) State, giving a reason, whether the initial rate of reaction would be the same as the average rate.

To find out the relationship between initial rate and the initial concentrations of the reactants, a series of experiments, in which the concentration of only one reactant is changed at a time, must be performed. The results of the initial concentrations and initial rates must then be compared.

The table below gives the experimental data for the reaction between propanone, CH_3COCH_3, and iodine, I_2, carried out in dilute hydrochloric acid.

$$CH_3COCH_3(aq) + I_2(aq) \longrightarrow CH_3COCH_2I(aq) + HI(aq)$$

Experiment	Initial concentrations / mol dm^{-3}			Initial rate / 10^{-4}
	I_2(aq)	CH_3COCH_3(aq)	HCl(aq)	mol dm^{-3} s^{-1}
1	0.0005	0.4	1.0	0.6
2	0.0010	0.4	1.0	0.6
3	0.0010	0.8	1.0	1.2

In experiments 1 and 2, only the concentration of iodine is changed and when it is doubled there is no change in the initial rate of reaction. Therefore the initial rate of reaction is independent of the initial concentration of aqueous iodine.

In experiments 2 and 3, only the concentration of propanone is changed and when it is doubled the initial rate of reaction also doubles.

Therefore the initial rate of reaction is directly proportional to the initial concentration of aqueous propanone.

Pointer

The amount of product formed in any reaction always depends on the amount of reactants even though the rate of reaction might not depend on the concentration of a particular reactant.

Factors affecting reaction rates

- Concentration of a solution (pressure of a gas)
- Surface area of a solid
- Temperature of a reaction
- Catalyst
- Light (in some reactions e.g. $H_2 + Cl_2$, photosynthesis).

How these factors change the rate of a reaction can be explained using collision theory.

Key Term

Activation energy is the minimum energy required to start a reaction by breaking of bonds.

Collision theory

For a chemical reaction to take place, reacting molecules must collide effectively. Only a small fraction of the total number of collisions leads to a reaction. However, the greater the number of collisions, the higher the chance that some of them will be effective. For a collision to be effective the molecules must collide in the correct orientation and have sufficient energy. Any factor that increases the chance of effective collisions will also increase the rate of reaction. The minimum energy needed is called the **activation energy**.

Pointer

The reactant species that collide during chemical reactions may include molecules, atoms or ions. For the sake of simplicity we refer to these simply as 'molecules'.

Energy profiles

The activation energy may be shown on diagrams called energy profiles. These compare the enthalpy of the reactants with the enthalpy of the products.

exothermic reaction

reaction pathway

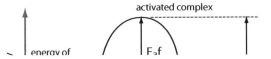

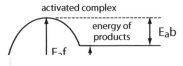

For an **exothermic reaction**, the reacting chemicals lose energy and heat is given out to the surroundings. Even though the products have a lower energy than the reactants, there still has to be an input of energy to break bonds and start the reaction.

For an **endothermic reaction** the enthalpy of the products is more than the enthalpy of the reactants and heat is taken in from the surroundings:

$\Delta H = E_a f - E_a b$

Where $E_a f$ and $E_a b$ are the activation energies of the forward and reverse reactions respectively.

For an exothermic reaction $E_a f < E_a b$ and ΔH is negative.

For an endothermic reaction $E_a f > E_a b$ and ΔH is positive.

Effect of concentration (pressure) on reaction rates

If the concentration of a reactant increases, the reaction rate increases. There are more molecules in the same volume so the distances between the molecules are reduced and there is an increase in the number of collisions per unit time. This means that there is a greater chance that there will be more collisions with energy greater than the activation energy hence the rate of reaction increases.

For a gaseous reaction, increasing the pressure is the same as increasing the concentration.

For a solid reducing the particle size increases the surface area and has the same effect.

>> *Pointer*

The activated complex may be regarded as a transition state in which old bonds have partly broken and new bonds have partly formed.

quickfire

② Consider the reaction:
$N_2 + 3H_2 \rightleftharpoons 2NH_3$
$\Delta H = -92$ kJ mol^{-1}.
  (a) Draw an energy profile diagram for the reaction.
  (b) Calculate the activation energy for the forward reaction given that the activation energy for the back reaction is 330 kJ mol^{-1}.

Grade boost

When explaining the effect of changing conditions on reaction rates always use collision theory in your answer. Bullet points can be useful.

Grade boost

When explaining the effect of temperature on reaction rate you must include a reference to activation energy in your answer.

Grade boost

When drawing a distribution curve remember to draw the line representing activation energy far to the right on the energy axis and ensure that the distribution curve does not touch the axis.

quickfire

③ Explain why animal products such as meat and milk stay fresher when refrigerated.

Key Term

A **catalyst** is a substance that increases the rate of a chemical reaction without being used up in the process. It increases the rate of reaction by providing an alternative route of lower activation energy.

Effect of temperature on reaction rates

If the temperature of a reaction increases, the reaction rate increases. At higher temperatures the molecules have more kinetic energy and are moving faster. More molecules have an energy that is greater than the activation energy and more collisions take place in a certain length of time. This can be shown using the Boltzmann energy distribution curves. In the diagram, temperature T_2 > temperature T_1.

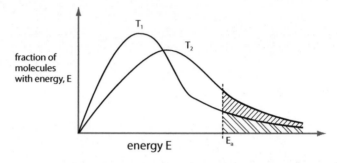

- The curves do not touch the energy axis.
- The areas under the two curves are equal and are proportional to the total number of molecules in the sample.
- At the higher temperature, T_2, the peak moves to the right (higher energy) with a lower height.
- Only the molecules with an energy equal to or greater than the activation energy, E_a, are able to react.
- At the higher temperature, T_2, many more molecules have sufficient energy to react and so the rate increases significantly.

Catalysts

A catalyst increases the rate of a chemical reaction without being used up in the process. A catalyst does take part in the reaction but can be recovered unchanged at the end of the reaction.

Catalysts work by providing a different reaction pathway for the reaction. The rate increases because the new pathway has a lower activation energy.

An energy profile diagram can be drawn to show the different pathways.

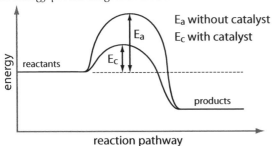

At the same temperature, a greater proportion of the reactant molecules will have sufficient energy to overcome the activation energy for a catalysed reaction. This can be shown on an energy distribution curve diagram.

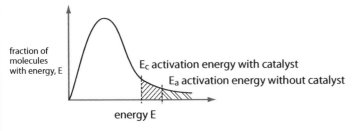

For a reversible reaction, a catalyst increases the rate of the forward and back reactions by the same amount, therefore it does not affect the position of equilibrium, but the position of equilibrium is reached more quickly.

There are two classes of catalysts: homogeneous and heterogeneous.

Homogeneous catalysts

A homogeneous catalyst is in the same phase as the reactants. Homogeneous catalysts take an active part in a reaction rather than being an inactive spectator. Examples are:

- Concentrated sulfuric acid in the formation of an ester from a carboxylic acid and an alcohol.
- Aqueous iron(II) ions, Fe^{2+}(aq), in the oxidation of iodide ions, I^-(aq), by peroxodisulfate(VI) ions, $S_2O_8^{2-}$(aq).

Heterogeneous catalysts

A heterogeneous catalyst is in a different phase from the reactants. Many industrially heterogeneous catalysts are d-block transition metals. The transition metal provides a reaction site for the reaction to take place. Gases are adsorbed on to the metal surface and react and the products desorb from the surface. The larger the surface area, the better the catalyst works. Examples are:

- Iron in the Haber process for ammonia production.
- Vanadium(V) oxide in the Contact process within sulfuric acid manufacture.
- Nickel in the hydrogenation of unsaturated oils in the production of margarine.
- Ziegler-Natta catalysts in the production of high density poly(ethene).

> **Pointer**
> A catalyst does not appear as a reactant in the overall equation of a reaction.

> **Key Terms**
> A **heterogeneous catalyst** is in a different phase from the reactants.
>
> A **homogeneous catalyst** is in the same phase as the reactants.

> **Grade boost**
> You should be able to state an example of a heterogeneous catalyst and a homogeneous catalyst in use.

Catalysts in industry

Most industrial processes involve catalysts. Industry relies on catalysts to reduce costs. Heterogeneous catalysts are often favoured in industry (see previous page for examples) because they are easily separated from the products. A catalyst speeds up a process by lowering the activation energy of the reaction so less energy is required for the molecules to react, and this saves energy costs. Much of this energy is taken from electricity supplies or by burning fossil fuel so a catalyst also has benefits for the environment. If less fossil fuel is burnt, less carbon dioxide will be released during energy production.

A motor company has developed a new catalyst for petrol-powered cars that needs only half the precious metals of current catalysts. The new catalyst uses nano-technology, which prevents the metals from clustering. The process of producing ethanoic acid from methanol originally used a pressure of 700 atm and a temperature of 300°C but by finding a new catalyst (iridium(IV) chloride) the process is now run at a pressure of 30 atm and a temperature of 150–200°C.

The biotechnology industry uses enzymes in a variety of important industrial processes such as food and drink production and the manufacturing of detergents and cleaners. Enzymes are biological catalysts found in all living things. They usually catalyse specific reactions and work best close to room temperature and pressure. Although enzymes are affected by temperature and pH and can be difficult to be removed from liquid products they are over 10 000 times more efficient than other catalysts in industry.

Some of the benefits are:

- Lower temperatures and pressures can be used, saving energy and costs.
- They operate in mild conditions and do not harm fabrics or food.
- They are biodegradable. Disposing of waste enzymes is no problem.
- They often allow reactions to take place which form pure products with no side reactions, removing the need for complex separation techniques.

quickfire

④ Give two reasons why the use of enzymes in industrial processes reduces the effect on the environment.

quickfire

⑤ Hydrogen peroxide solution decomposes slowly to water and oxygen at room temperature. Give four ways in which the rate of decomposition can be speeded up.

Studying rates of reaction

To measure the rate of a chemical reaction we need to find a physical or chemical quantity which varies with time. These are some of the methods (all carried out at constant temperature).

- **Change in gas volume**   e.g. $Mg(s) + 2HCl(aq) \rightarrow MgCl_2(aq) + H_2(g)$

  In a reaction in which gas is formed, the volume of the gas can be recorded using a gas syringe at various times.

- **Change in gas pressure**   e.g. $PCl_5(g) \rightarrow PCl_3(g) + Cl_2(g)$

  Some reactions between gases involve a change in the number of moles of gas. The change in pressure (at constant volume) at various times can be followed using a manometer.

- **Change in mass**   e.g. $CaCO_3(s) + 2HCl(aq) \rightarrow CaCl_2(aq) + H_2O(l) + CO_2(g)$

  If a gas forms in a reaction and is allowed to escape, the change in mass at various times can be followed using a weighing balance.

- **Change in colour (colorimetry)**

  Some reaction mixtures show a steady change of colour as the reaction proceeds. The concentration of the substance changing colour can be monitored using a colorimeter.

  e.g. $CH_3COCH_3(aq) + I_2(aq) \rightarrow CH_3COCH_2I(aq) + HI(aq)$

  The intensity of the colour of the iodine can be monitored over time, since iodine (brown) is the only coloured species in the reaction, using a colorimeter and hence its change in concentration can be measured.

Specified practical work

There are three practical exercises in this topic.

Gas collection method

This is a good way of showing how rate changes during a chemical reaction as well as illustrating how changing concentration, temperature, particle size or catalysts can affect a chemical reaction.

A suitable example is reacting a metal with an acid. A typical method would be:

- Start the reaction by shaking the metal into the acid and start a stopwatch.
- Measure the amount of hydrogen given off at constant intervals.
- Stop the watch when hydrogen is no longer being produced.
- Repeat the experiment with different concentrations of acid / temperature of acid / particle size of metal ensuring that all other factors are kept constant.
- Draw a graph of your results.

quickfire

⑥ Dinitrogen pentoxide decomposes according to the equation

$$2N_2O_5(g) \rightarrow$$
colourless
$$4NO_2(g) + O_2(g)$$
brown   colourless

Describe two methods by which the rate of this reaction can be measured.

≫ Pointer

Other suitable examples for a gas collection method are: Addition of calcium carbonate to hydrochloric acid. (Sulfuric acid is not suitable.) Decomposition of hydrogen peroxide. (Different catalysts can be compared.) Instead of using a gas syringe, the gas can be collected over water using an inverted burette.

quickfire

⑦ State why the rate of reaction between $CaCO_3$ and H_2SO_4 cannot be followed using a gas collection method.

Iodine-clock reactions

To compare rates of reaction under different conditions, a number of experiments may be set up in which initial concentrations of reactants are known and the time taken for each experiment recorded.

Iodide ions can be oxidised to iodine at a measurable rate. Iodine gives a strongly coloured blue complex with starch solution but if a given amount of thiosulfate ion – with which iodine reacts very rapidly – is added, no blue colour will appear until enough iodine has been formed to react with all the thiosulfate. The time taken for this to occur thus acts as a 'clock' to measure the rate of iodide ions being oxidised.

Oxidising iodide ions by hydrogen peroxide in acid solution is a suitable example of an 'iodine clock' reaction.

$$H_2O_2(aq) + 2H^+(aq) + 2I^-(aq) \xrightarrow{\text{slow}} 2H_2O(l) + I_2(aq)$$

$$I_2(aq) + 2S_2O_3^{2-}(aq) \xrightarrow{\text{fast}} 2I^-(aq) + S_4O_6^{2-}(aq)$$

The temperature must be kept constant, since rates vary rapidly with changes in temperature.

A typical method for this reaction would be:

- Accurately measure known volumes of acid, thiosulfate solution and iodide solution into a conical flask and add a little starch solution.

- Accurately measure a known volume of hydrogen peroxide into a test-tube.

- Rapidly pour the peroxide into the flask, simultaneously start a stopwatch and mix thoroughly.

- When the blue colour appears stop the watch.

- Repeat using five different concentrations of peroxide, ensuring that the total volume of the mixture is constant.

- The concentration of peroxide should vary by at least threefold to ensure a good spread of results.

Since rate $\propto \dfrac{1}{\text{time}}$ and total volume is constant in each case, $[H_2O_2] \propto$ volume of peroxide used in each run. Plotting a graph of $\dfrac{1}{\text{time}}$ against volume of peroxide will give the relationship between $[H_2O_2]$ and rate.

A similar procedure can be used varying the concentration of the potassium iodide to find the effect of $[I^-]$ on reaction rate. However, the peroxide solution must be added last every time.

≫ Pointer

In the reaction between iodide ions and hydrogen peroxide in acid solution, a large batch containing acid, iodide, thiosulfate and starch in the correct proportions can be made up if preferred. When investigating the effect of peroxide concentration, a known volume of this mixture can be added to a conical flask, using a burette, each time a different concentration of peroxide solution is used. Another suitable example is: Oxidation of iodide ions by peroxodisulfate ions to form iodine.

Add a fixed volume of aqueous thiosulfate ions and a few drops of starch solution to the reaction mixture. The thiosulfate rapidly reduces the iodine formed. When all the thiosulfate has reacted, free iodine quickly forms a blue complex with the starch.

Precipitation reactions

It is not only colour changes in solutions that can be used to follow rates of reaction. Sometimes colourless solutions become more and more cloudy as a precipitate forms and the time taken for a precipitate to form can be used to measure the rate of a reaction.

An example of this is the reaction of thiosulfate ions in acid solution.

$$S_2O_3^{2-}(aq) + 2H^+(aq) \longrightarrow S(s) + H_2O(l) + SO_2(g)$$

The reaction is easy to follow since one sulfur atom is formed for each thiosulfate ion reacting and the sulfur makes the reacting solution more cloudy as its concentration increases. By placing the reaction vessel over a black cross (which will disappear from view when a fixed amount of reaction has produced a fixed amount of sulfur) the rates of reaction of solutions of differing concentrations can be compared and the effects of changing concentrations on the rate found.

This is because the time taken for this fixed amount of reaction in all runs is inversely proportional to the rate of reaction, i.e. if the reaction is fast the cross will disappear quickly and vice versa.

The temperature must be kept constant, since rates vary rapidly with changes in temperature.

A typical method for this reaction would be:

- Accurately measure a known volume of thiosulfate solution into a conical flask.
- Accurately measure nitric acid into a test tube.
- Rapidly pour the acid into the flask, simultaneously start a stopwatch and mix thoroughly.
- Place the flask over a black cross and stop the watch the moment the black cross can no longer be seen.
- Repeat the procedure with three different concentrations of thiosulfate keeping the volume of acid constant and the total volume of the mixture constant.
- Repeat the procedure with three different concentrations of acid keeping the volume of thiosulfate constant and the total volume of the mixture constant. (The concentration of each should vary by at least threefold to ensure a good spread of results.)

Since rate $\propto \dfrac{1}{time}$ plotting a graph of $\dfrac{1}{time}$ against concentration of thiosulfate at constant acid concentration will give the relationship between $[S_2O_3^{2-}]$ and rate. Plotting a graph of $\dfrac{1}{time}$ against concentration of acid at constant thiosulfate concentration will give the relationship between $[HNO_3]$ and rate.

 Pointer

In an examination, you will be expected to analyse given experimental data to follow the rate of a reaction.

2.3 The wider aspects of chemistry

(a) Application of the principles studied to problems encountered in the production of chemicals and of energy

Pointer

When tackling trial questions, refer continually to the relevant outcomes in 2.1 and 2.2.

This topic requires you to apply the principles you have studied in Components 2.1 and 2.2 to situations and problems encountered in the production of chemicals and energy. You may be supplied with data relevant to the situation, process or problem which you may not have met before and will be marked on your ability to analyse and evaluate the situation or problem, usually by answering a series of questions on the topic. Calculations may be required and clearly a basic understanding of equilibrium, energetics and kinetics will be important in tackling many of the questions.

There are no specific learning outcomes as such, what is needed is some practice and experience in applying the relevant sections of 2.1 and 2.2 to the particular question. Here a study of recent questions in past Component 1 papers on the WJEC website (www.wjec.co.uk) may be useful:

Q8 in January 2010 dealt with the chemical removal of carbon dioxide from power station flues, the associated equilibria and temperature effects and a direct separation using a membrane formed by nanotechnology.

Q8 in January 2011 required an evaluation of the carbon dioxide concentration/global temperature relation.

Grade boost

The questions will be fairly straightforward but may need some flexibility and lateral thinking to see what is needed in what may be an unfamiliar problem.

quickfire

① Which three topics from 2.1 and 2.2 will be useful when answering a question about the effects of changing temperature, pressure and catalyst on the yield and rate of production of polyethene?

Energy

The main problem is the huge and increasing amounts of energy required that is mostly provided by the combustion of non-renewable fossil fuels with the associated release of carbon dioxide into the atmosphere and oceans. Atmospheric levels have increased by around one-third in the last 100 years with a probable contribution to global warming, and the oceans are becoming more acidic with serious biological effects. At present consumption rates, the supply of fossil fuels will diminish and their cost increase so that much work is being done on alternative sources; those of chemical interest include nuclear power, solar energy and biomass fuels, the aim of the latter being to achieve **carbon neutrality.** In this, the CO_2 generated in combustion of the biomass is compensated for by that absorbed during the photosynthetic growth of the plants.

Grade boost

Make it very clear when describing atom economy that this is based on **mass** or relative molecular mass and not on the number of molecules in the equation. Do not use the word 'amount', since this is connected via 'amount of substance' to the number of moles.

quickfire

② Define the term percentage atom economy.

Nuclear power is well established, contributing around 20% of fuel needs, and efforts now focus on making it safer. Solar power uses semiconductor chemistry to obtain increased efficiency and reliability and there is also interest in using hydrogen as a fuel since no CO_2 is produced during its combustion. However, hydrogen does not occur naturally on Earth and efforts are being made to form it through the solar photolysis of water.

(b) The role of Green Chemistry and the impact of chemical processes

NB Under these topics you may be supplied with data and information as a basis for the discussion and evaluation in your answer.

The aim of Green Chemistry is to make the chemicals and products that we need with as little impact on the environment as possible. This means:

- Using as little energy as possible and getting this from renewable sources such as biomass, solar, wind and water rather than from finite fossil fuels such as oil, gas and coal. Generally increase energy efficiency.

- Using renewable raw materials (feedstocks) such as plant-based compounds whenever possible.

- Using methods having high atom economy so that a high percentage of the mass of reactants ends up in the product, giving little waste, see page 31 for Atom economy.

- Developing better catalysts, such as enzymes, to carry out reactions at lower temperatures and pressures to save energy and avoiding constructing strong plants

- Avoid using solvents, especially volatile organic solvents that are bad for the environment.

- Make products that are biodegradable at the end of their useful lives, where possible.

- Avoid using toxic materials if possible and ensure that no undesirable co-products or by-products are released into the environment to prevent pollution.

Impacts

The chemical industry plays a major role in modern life as a producer of the materials we need, from semiconductors in mobile phones to fertilisers in agriculture. It is a large producer of economic wealth and a major employer. While occasional accidents get wide publicity, most operations are clean and safe, carefully controlled and regulated and located away from centres of population.

The major chemical reaction today is the combustion of fossil fuels in vehicles, homes and factories, not to produce a product but to provide energy. See 2.3(a) above. The unwanted carbon dioxide product is giving an accelerating increase in atmospheric levels, and probably to an increase in global warming. To minimise this is a major research area in chemistry and other sciences and one aim of Green Chemistry, as above.

a. Hydrogen gas is a possible combustion fuel for the future ($\Delta H = -286$ kJ mol^{-1}).

  State one advantage and one disadvantage of using this.

b. Coffee beans are now mainly decaffeinated using supercritical carbon dioxide under pressure as against the former method of using an organic solvent.

  Suggest one advantage and one disadvantage of using CO_2.

c. 1,2-Dichloroethane is a precursor in PVC manufacture and may be made by two catalysed reactions:

  (i) Ethene + chlorine = CH_2ClCH_2Cl ($\Delta H = -185$ kJ mol^{-1})

    No added solvent or heating needed, 40° operating temp.

  (ii) Ethene + HCl + O_2 = CH_2ClCH_2Cl + H_2O ($\Delta H = -242$ kJ mol^{-1})

    300° operating temp. and 5 atm. pressure.

  Compare the two processes and consider their advantages and disadvantages. Yields are similar for the two.

2.4 Organic compounds

Key Terms

Hydrocarbon is a compound of carbon and hydrogen only.

Functional group is the atom/group of atoms that gives the compound its characteristic properties.

Saturated compound is one that contains no C to C multiple bonds.

Unsaturated compound is one that contains C to C multiple bonds.

Grade boost

The longest C chain can be bent/go round corners.

quickfire

① Find and number the longest C chain in the compound below.

What is the 'code' that applies to the name of this compound?

Historically organic compounds were derived from living species but nowadays the term is applied to most carbon-containing compounds.

Naming organic compounds

There are millions of possible organic compounds and each one has a specific name. To name a compound you have to know the homologous series to which it belongs – generally the **functional group** it contains.

The homologous series in this unit are:

alkanes – saturated hydrocarbons

alkenes – unsaturated hydrocarbons with a C to C double bond

halogenoalkanes – compounds in which one or more hydrogens in an alkane have been replaced by a halogen

primary (1°) alcohols – compounds containing –OH as the functional group

carboxylic acids – compounds containing the $-\overset{\overset{\text{O}}{\|}}{\text{C}}-\text{OH}$ (sometimes written as $-COOH$ or $-CO_2H$) as the functional group.

You also need to know the 'code' that applies to the number of carbon atoms. In this:

meth	= 1 carbon	pent	= 5 carbons	non	= 9 carbons
eth	= 2 carbons	hex	= 6 carbons	dec	= 10 carbons.
prop	= 3 carbons	hept	= 7 carbons		
but	= 4 carbons	oct	= 8 carbons		

Rules

1 Find the longest continuous carbon chain. Use the code above as the basis of the name.

2 Number the C atoms in the chain, starting from the end that gives any side chains or substituted groups the smallest numbers possible.

3 If there is more than one side chain or substituted group the same, use the prefix di for 2, tri for 3 and tetra for 4.

4 Keep the alphabetical order of branch name.

It is easier to see how the rules work by applying them to particular examples.

2-methyl pentane   2,2-dimethyl propane

Note: the – CH_3 group is called methyl.

1-bromo, 4-methyl, hex-1-ene, 3-ol

Br   H   OH   H   H

(structure with six-carbon chain, C=C, OH group, Br, and methyl branch)

Note: – OH is the functional group of alcohols and therefore the name includes – ol.

C=C is the functional group for alkenes and therefore the name includes –ene.

Bromo shows the presence of bromine.

There are six carbons in the longest carbon chain – you count along the chain even if it is bent.

Types of formulae

The formula of a particular compound can be shown in several different ways.

Molecular formula: shows the atoms, and how many of each type there are, in a molecule of compound.

Displayed formula: shows all the bonds and atoms in the molecule.

Shortened formula: shows the groups in sufficient detail that the structure is unambiguous.

Skeletal formula: shows the carbon/hydrogen backbone of the molecule as a series of bonds with any functional groups attached.

The different types can be seen using 4-bromo, hex-2-ene

Molecular formula: $C_6H_{11}Br$

Displayed formula:

(displayed formula structure with H, Br substituents)

Shortened formula: $CH_3CHCHCHBrCH_2CH_3$

Skeletal formula:
Br

Homologous series

As described earlier, each set of compounds considered above belongs to a particular homologous series. An homologous series is a set of compounds that:

1 can all be represented by a general formula;

2 differ from their neighbour in the series by CH_2;

3 have the same functional group and so very similar chemical properties;

4 have physical properties that vary as the M_r of the compound varies.

The general formula of the alkane homologous series is C_nH_{2n+2} where n is an integer.

The effects of the functional group and the way in which physical properties, particularly melting and boiling temperatures, vary within an homologous series are considered later.

quickᖴire

② Draw the formulae of:
  (a) 3- bromo pent-2-ene
  (b) 2.2- dimethyl hexan-3-ol

quickᖴire

③ What are the names of the following?
  (a)

  (b) (structure with H, OH, OH, H, Br substituents)

Grade boost

In an exam make sure you use the type of formula that the question asks for – molecular, displayed, shortened or skeletal.
There could also be empirical formulae – these are covered later!

quickᖴire

④ The shortened formula of a compound is shown below.
  $CH_2=CHCH_2CHClCH_3$
  (a) What is its molecular formula?
  (b) Draw its displayed formula.
  (c) Draw its skeletal formula.

Grade boost

In a skeletal formula carbon atoms are never shown and only hydrogen atoms within the functional group are shown.

Key Term

Empirical formula is the formula of a compound with the atoms of the elements in their simplest ratio.

quickfire

⑤ Name the compound whose skeletal formula is shown below:

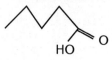

Grade boost

Always show clearly that you have calculated the M_r for your empirical formula.

quickfire

⑥ What is the general formula of the alkene homologous series?

quickfire

⑦ What is the molecular formula of the alkane with 72 carbons?

quickfire

⑧ A molecule has M_r of approximately 188. Its percentage composition is:
C = 12.78; H = 2.15; Br = 85.07.
Find its empirical and hence its molecular formula.

Grade boost

Do not approximate when you 'divide by the smallest'. 1.67, for example, is $1\,^2/_3$. In this case you should multiply by 3 (to give 5).

Empirical formulae

Methods used to analyse organic compounds often give results as masses or percentages of the elements present. These can be used to determine the empirical formula.

Example using masses

0.205 g of a hydrocarbon was burnt completely in oxygen and 0.660 g of carbon dioxide and 0.225 g of water were formed. The M_r of the compound was approximately 80.

(i) Calculate the empirical formula.

Mass of carbon in CO_2 = 12/44 × 0.660 = 0.180 g

Mass of hydrogen in H_2O = 2/18 × 0.225 = 0.025 g

$$\text{Ratio of number of moles} \quad C : H$$

$$= \quad \frac{0.180}{12.0} : \frac{0.025}{1.01}$$

$$= \quad 0.0150 : 0.0248$$

$$\text{Divide by smaller} \quad = \quad 1 : 1.65$$

$$= \quad 3 : 5$$

Empirical formula is C_3H_5

(ii) What is the molecular formula of the hydrocarbon?

Since the molecular formula can be the same as, or a multiple of, the empirical formula, it is necessary to use its M_r.

M_r of empirical formula = 41

M_r of compound approximately 80 and therefore molecular formula is C_6H_{10}

Example using percentages

Analysis of a compound gave the percentage composition C = 60.0%, H = 13.3%, O = 26.7%. The M_r of the compound was 60. Determine its molecular formula.

$$\text{Ratio} \quad C : H : O$$

$$= \quad \frac{60.0}{12.0} : \frac{13.3}{1.0} : \frac{26.7}{16.0}$$

$$= \quad 5.00 : 13.3 : 1.67$$

$$= \quad 3 : 8 : 1$$

Empirical formula = C_3H_8O

Empirical M_r = 60, therefore molecular formula = C_3H_8O.

Isomerism

Structural

Structural isomerism can arise in several ways:

Chain isomerism when the carbon chain is arranged differently.

Example

$CH_3CH_2CH_2CH_2CH_3$ pentane

and

$$CH_3 - \overset{\overset{\displaystyle CH_3}{|}}{\underset{\underset{\displaystyle H}{|}}{C}} - \overset{\overset{\displaystyle H}{|}}{\underset{\underset{\displaystyle H}{|}}{C}} - CH_3 \quad \text{2-methylbutane}$$

and

$$CH_3 - \overset{\overset{\displaystyle CH_3}{|}}{\underset{\underset{\displaystyle CH_3}{|}}{C}} - CH_3 \quad \text{2,2-dimethylpropane}$$

Position isomerism when the functional group is in a different position.

Example

$CH_2ClCH_2CH_3$ 1-chloropropane

$CH_3CHClCH_3$ 2-chloropropane

Functional group isomerism when the functional group is different.

Example

$CH_3CH_2OCH_3$ an ether

$CH_3CH_2CH_2OH$ propan-1-ol

Key Term

Structural isomers are compounds with the same molecular formula but a different structural formula, i.e. arrangement of the atoms.

Grade boost

You will be expected to recognise, name and draw isomers but those involving different functional groups are considered in a later unit.

quickfire

⑨ Draw displayed formulae for three of the structural isomers of hexane. Name the isomers.

quickfire

⑩ Draw skeletal formulae for two structural isomers of hexene that involve different positions of the double bond. Name the isomers.

Key Term

E-Z isomerism occurs in alkenes (and substituted alkenes) due to restricted rotation about the double bond.

Grade boost

In older text books you may see reference to *cis-trans* isomerism. This has now been superseded by *E-Z*.

quickfire

⑪ Draw the Z isomer of 1-iodo 2-bromoethene.

quickfire

⑫ Does the structure below show the *E* or *Z* isomer?

Br   CH₃
   \\   /
    C = C
   /   \\
H   Cl

Grade boost

If you need a mnemonic to remember which is which, *Z* comes from the German *zusammen* = together and *E* from *entgegen* = opposite.

E-Z isomerism

The single bond in alkanes allows free rotation but the double bond in alkenes means that rotation is restricted. This is due to the π bond – the reason for this is described later.

Since the double bond does not rotate, compounds such as 1,2-dichloroethene can exist in two different forms, i.e. two different isomers.

Cl   Cl   Cl   H
  \\   /   \\   /
    C = C   C = C
  /   \\   /   \\
H   H   H   Cl
      (Z)   (E)

Naming *E-Z* isomers

To decide which isomer is which, you must look at the nature of the atom directly attached to each of the carbon atoms in the double bond. Look first at each carbon separately. The atom with the higher atomic number takes precedence.

Then look at how the higher priority groups are arranged – if they are both on the same side of the double bond, it is the *Z* isomer and if they are on opposite sides, it is the *E* isomer.

Examples

For the isomers of 1,2-dichloroethene shown above:

on each side of the double bond Cl has the higher A_r and therefore precedence;

in the LH structure the chlorines are on the same side of the double bond, i.e. this is the Z form.

If two different groups are attached across the double bond the groups around each carbon must be considered separately.

I   H
  \\   /
    C1 = C2
  /   \\
H   Cl

Looking at C 1, iodine has priority whilst at C 2 chlorine has priority. The atoms with higher priority are on opposite sides of the double bond so that this is (E) – 1-chloro 2-iodoethene.

Properties of *E-Z* isomers

Since the functional groups are held differently, *E-Z* isomers can have different physical and chemical properties.

In, for example (Z) and (E) -butenedioic acids, the 2 acid groups can interact with each other in the Z form but are too far apart in the E form – this affects chemical reactions.

One example is the dehydration of (Z) – butenedioic acid.

$+H_2O$

In the E form the COOH groups are far apart and this reaction cannot occur.

They will also pack together differently and this affects physical properties – particularly melting and boiling temperatures. In general the E form packs together better so that it has stronger intermolecular forces and higher melting and boiling temperatures.

Melting and boiling temperatures of organic compounds

The effect of chain length on melting and boiling temperatures

When a simple covalent substance melts or boils, heat energy is supplied. This is to overcome **Van der Waals forces**. Which substance has a higher melting or boiling temperature can be predicted by looking at the strength of these forces.

For hydrocarbons only induced dipole–induced dipoles are present and these are weak. Since the intermolecular forces happen at the surface, it is necessary to look at the surface areas of the molecules. A small surface area means that lower Van der Waals forces are possible and this means that the melting and boiling temperature would be low.

Small hydrocarbons are therefore gases at room temperature, larger ones are liquids and even larger ones are solids.

You know that the boiling temperature of compounds increases with increasing chain length. If different structural isomers are considered, they have different surface areas. This can be used to explain boiling temperatures.

The more branches an isomer has, the more like a sphere it is and the lower is the boiling temperature.

many branches – little surface area contact

straight chain – more surface area contact

Key Term

Van der Waals forces = dipole–dipole or induced dipole–induced dipole interactions between atoms and molecules.

Grade boost

Look back at the section on Van der Waals forces and make sure you can explain the formation of induced dipole–induced dipole forces.

quickfire

⑬ The boiling temperature of hexane is 68°C. Suggest a value for the boiling temperature of heptane.

quickfire

⑭ Draw structural formulae for pentane and 2,2-dimethylpropane and use these to explain why the boiling temperature of pentane is 36°C whilst that of 2,2-dimethylpropane is 10°C.

Key Terms

Fossil fuel is one that is derived from organisms that lived long ago.

Non-renewable resources are those that cannot be reformed in a reasonable timescale.

Greenhouse gas is one that causes an increase in the Earth's temperature.

Acid rain is rain with lower than expected pH.

Complete combustion is combustion that occurs with excess oxygen.

Incomplete combustion is combustion that occurs with insufficient oxygen.

Grade boost

Answers to questions often need discussion of both advantages and disadvantages of the use of fossil fuels.

quickfire

① (a) Name two gases that cause acid rain.

(b) Describe a source for each of these gases.

Grade boost

Carbon monoxide is toxic/poisonous. Be precise when describing chemical hazards – 'dangerous' is too vague.

quickfire

② Why do cars that give black smoke in their exhaust not have good fuel economy?

quickfire

③ What is the formula of an alkane with 90 carbon atoms?

2.5 Hydrocarbons

Fossil fuels

Historically we have been dependent on **fossil fuels** as our energy source and, although other sources are being developed, this dependence seems set to continue for the foreseeable future.

The use of fossil fuels has both advantages and disadvantages.

Advantages

1 The variety of fuels available, from natural gas through to coal, means that each use can be matched to an appropriate form of fuel.

2 Fossil fuels are available at all times – unlike, for example, wind and solar.

Disadvantages

1 They are **non-renewable** since their formation takes millions of years and reserves are being used more rapidly than new ones are formed.

2 The combustion of fossil fuels produces the **greenhouse gas** carbon dioxide. The associated rise in temperatures has serious environmental consequences including rising sea levels and changes in the crops that can be grown in particular places.

3 The combustion of fossil fuels containing sulfur produces sulfur dioxide whilst internal combustion engines produce oxides of nitrogen. These can react with rainwater to give **acid rain** containing sulfuric acid and nitric acid.

This can lead to environmental problems such as damage to buildings, vegetation and aquatic life as well as health issues for people with breathing difficulties.

4 Carbon monoxide is formed when **incomplete combustion** of fossil fuels occurs. This is highly toxic.

Alkanes

The homologous series of alkanes

1 The general formula of the alkane homologous series in C_nH_{2n+2}.

2 Each member of the series differs from Its neighbour by CH_2.

3 The chemical properties of all the series are similar.

4 The physical properties vary gradually as the relative formula mass increases. Small alkanes are gases (natural gas is mainly methane), larger alkanes are liquids (petrol has approximately 8 carbon atoms in a molecule) and even larger alkanes are solids (wax candles).

Reactions of alkanes

Alkanes are non-polar and do not contain multiple bonds. This means that they are generally unreactive. There are, however, two important reactions of alkanes.

1 Combustion

Alkanes burn and react with oxygen in exothermic reactions and are therefore used as fuels.

If sufficient oxygen is present complete combustion occurs to produce carbon dioxide and water.

Using propane:

$$C_3H_8(g) + 5O_2(g) \longrightarrow 3CO_2(g) + 4H_2O(l)$$

If insufficient oxygen is present, incomplete combustion occurs and carbon monoxide is formed. This is toxic since it can inhibit transport of oxygen through the body. It also produces less energy than complete combustion.

Incomplete combustion also produces carbon and this is responsible for the black smoke that can be seen when diesel engines are not properly adjusted.

2 Halogenation

Halogens will react with alkanes in the presence of uv light (often from the sun).

The mechanism of this reaction is in three stages.

Key Terms

Halogenation is a reaction with any halogen.

Initiation is the reaction that starts the process.

Homolytic bond fission is when a bond is broken and each of the bonded atoms receives one of the bonded electrons.

Radical is a species with an unpaired electron.

Propagation is the reaction by which the process continues/grows.

Chain reaction is one that involves a series of steps and, once started, continues.

Termination is the reaction that ends the process.

A **reaction mechanism** shows the stages by which the reaction proceeds.

A **substitution reaction** is one in which one atom/ group is replaced by another atom/ group.

quickfire

④ Why are alkanes generally so unreactive?

Grade boost

There are only three mechanisms that you need to be able to describe. At least one of them is asked on nearly every exam and so you should make sure you really understand them!

quickfire

⑤ Why are radicals so reactive?

quickfire

⑥ How is C_2H_6 formed as one product in the reaction between methane and chlorine?

quickfire

⑦ Write the overall equation for the formation of $CHCl_3$ from CH_4.

Grade boost

The stages (initiation, etc.) are the mechanism. A question may ask for this or the overall equation such as:

$C_2H_6 + Cl_2 \rightarrow C_2H_5Cl + HCl$

The mechanism of this reaction is in three stages.

Stage 1 – initiation

The energy needed to break the bond in chlorine is provided by uv light. This is **homolytic bond fission**.

$$Cl_2 \rightarrow 2Cl^{\bullet}$$

$Cl^{\bullet}$ is then a **radical**.

Stage 2 – propagation

Radicals are very reactive and will take part in a series of **propagation** reactions.

$$Cl^{\bullet} + CH_4 \rightarrow CH_3{}^{\bullet} + HCl$$
$$CH_3{}^{\bullet} + Cl_2 \rightarrow CH_3Cl + Cl^{\bullet}$$

In the propagation stages each reaction starts with a radical and then produces one so that the **chain reaction** continues.

Stage 3 – termination

The chain reaction continues until two radicals meet in a **termination** stage.

$$CH_3{}^{\bullet} + Cl^{\bullet} \rightarrow CH_3Cl$$

Further substitution

Since radicals are so reactive other propagation stages can occur, e.g.

$$CH_3Cl + Cl^{\bullet} \rightarrow CH_2Cl^{\bullet} + HCl$$
$$CH_2Cl^{\bullet} + Cl_2 \rightarrow CH_2Cl_2 + Cl^{\bullet}$$

This means that polysubstitution can occur and a mixture of products is formed. This means that it is not really a satisfactory method to prepare chloromethane although this product will dominate if the amount of halogen is limited.

Overall, this type of mechanism is described as being radical substitution.

Alkenes

The alkenes are an homologous series of unsaturated hydrocarbons. The functional group is a carbon to carbon double bond and the general formula of the series in C_nH_{2n}.

Alkenes are formed when long chain hydrocarbons are cracked and they are used in the production of polymers and as starting materials for the preparation of a range of organic compounds.

Since alkenes contain a double bond they are much more reactive than alkanes.

Structure and bonding

Alkenes contain a carbon to carbon double bond. In ethene the double bond consists of a sigma (σ) bond and a pi (π) bond. The π bond is formed from the sideways overlap of a p orbital electron on each carbon.

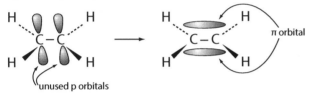

Reactions

The mechanism of most reactions of alkenes is **electrophilic addition.**

The pair of electrons in the π orbital means that alkenes have a region of high electron density and are susceptible to attack by an **electrophile**. The mechanism involves **heterolytic bond fission** and leads overall to electrophilic **addition**.

If the attacking species is not polar, a dipole is induced by the negative charge of the π bond.

X_2 could be H_2 or Br_2.

Key Terms

A π **bond** is one formed by the sideways overlap of p electrons.

Electrophile is an electron deficient species that can accept a lone pair of electrons.

Heterolytic bond fission is when a bond is broken and one of the bonded atoms receives both the bonded electrons.

Addition reaction is a reaction in which reagents combine to give one product.

Grade boost

When drawing a mechanism you must show exactly where the 'curly arrows' start and finish. The arrow generally starts from a lone pair or a π bond.

quickpire

⑧ Why is the mechanism of addition to alkenes said to involve heterolytic bond fission?

Grade boost

Be careful when to use + and when to use $\delta+$. + is produced when a neutral species loses an electron, $\delta+$ forms as part of a dipole.

Key Term

Carbocation (carbonium ion) is a positively charged carbon-containing species.

Grade boost

In describing a test, always include the appearance before and after the test.

Grade boost

Potassium manganate(VII) is an oxidising agent that can oxidise other organic compounds. Alkenes decolorise purple manganate (VII) ions but so do other compounds.

quickfire

⑨ What is the effect, on the melting temperature, of adding hydrogen to polyunsaturated oils?

quickfire

⑩ Complete the equation:
$(CH_3)_2C=CH_2 + HBr \rightarrow$

Examples of the addition reaction

(i)   Test for alkenes

The reaction with bromine is used as a test for alkenes since the brown colour of bromine changes to colourless. In practice aqueous bromine (bromine water) is generally preferred since it is safer.

The reaction with potassium manganate(VII) can also be used as a test for the presence of a carbon to carbon double bond. The purple manganate(VII) Is decolorised as two OH groups add across the double bond to form a diol.

(ii) Reaction with hydrogen

This can be catalysed by several transition elements but nickel is most commonly used. The reaction is called hydrogenation and it is commercially important since it converts liquid polyunsaturated oils into solid, more saturated edible fats. These are used as spreads/ butter substitutes.

(iii) Reaction with hydrogen bromide

Unsymmetrical alkenes could add HBr to give two different products.

$$
\begin{array}{ll}
\underset{\text{propene}}{\ce{CH2=CH-CH3}} + HBr & \longrightarrow \underset{\text{2-bromopropane}}{\ce{CH3-CHBr-CH3}}
\end{array}
$$

propene   2-bromopropane

$$
\begin{array}{ll}
\underset{\text{propene}}{\ce{CH2=CH-CH3}} + HBr & \longrightarrow \underset{\text{1-bromopropane}}{\ce{CH2Br-CH2-CH3}}
\end{array}
$$

propene   1-bromopropane

The major product is 2-bromopropane due to the greater stability of the 2° **carbocation** (compared with the 1° carbocation).

2° carbocation   1° carbocation

Polymerisation

Polymerisation is the joining of a very large number of **monomer** molecules to make a large polymer molecule.

Alkenes, and substituted alkenes, undergo **addition polymerisation**. In this type of polymerisation the double bond is used to join the monomers and nothing is eliminated.

Ethene is polymerised to make poly(ethene) (commonly called polythene).

The names of **polymers** still include the name of the monomer but the polymers do not contain double bonds.

Poly(ethene) is unreactive and flexible so it is used to make plastic bags, etc.

When poly(ethene) was first made, the polymer chains had side branches coming from the main chain and these prevented the polymer chains packing together. This meant that the density of the polymer was low and there were few points of contact for Van der Waals forces so that the melting temperature was also low.

Catalysts can be used to make poly(ethene) with straight chains. This means that the chains can pack together so that the polymer has a higher density and higher melting temperature. These properties mean that it is used where more rigidity is needed and/or the temperature is comparatively high.

The properties of polymers can also be altered by using substituted alkenes as the monomer. This means that these polymers have a huge variety of uses.

Note: You do not need to be able to quote specific uses for specific polymers but you should be aware of the principles involved in the polymerisation and how the different physical properties make different polymers suitable for different functions.

Key Terms

Polymerisation Is the joining of a very large number of monomer molecules to make a large polymer molecule.

Monomer is a small molecule that can be made into a polymer.

Polymer is a large molecule made by joining many monomer molecules.

Repeat unit is the section of the polymer that is repeated to make the whole structure

Grade boost

When drawing a section of a polymer chain you must show – at both ends to show that the chain continues.
If you only draw one **repeat unit**, do not forget to put 'n' outside the bracket.

quickfire

⑪ What is the name of the polymer formed from 1-chloro, 2-cyanoethene?

quickfire

⑫ Draw two repeat units for the polymer you have named above. (The cyano group is CN.)

quickfire

⑬ What is the empirical formula of poly(ethene)?

Grade boost

An easy way of drawing the formula of a polymer when you are told the monomer is to draw the monomer with all substituted groups around the double bond i.e.

$$\overset{W}{\underset{Y}{>}}C=C\overset{X}{\underset{Z}{<}}$$

Then just break the double bond.

Other widely used polymers

These include:

Poly(propene)

$$n \quad \overset{CH_3}{\underset{H}{>}}C=C\overset{H}{\underset{H}{<}} \longrightarrow \left[\begin{array}{cc} CH_3 & H \\ | & | \\ C & - C \\ | & | \\ H & H \end{array}\right]_n$$

monomer
propene

polymer
poly(propene)

Poly(propene) is rigid and used in food containers and kitchen equipment.

quickfire

⑭ A section of polymer is shown below. Draw the monomer that was used to form this polymer.

$$-\overset{Cl}{\underset{H}{\overset{|}{C}}} - \overset{OH}{\underset{H}{\overset{|}{C}}} - \overset{H}{\underset{Cl}{\overset{|}{C}}} - \overset{H}{\underset{OH}{\overset{|}{C}}} -$$

Poly(chloroethene)

$$n \quad \overset{H}{\underset{H}{>}}C=C\overset{Cl}{\underset{H}{<}} \longrightarrow \left[\begin{array}{cc} H & Cl \\ | & | \\ C & - C \\ | & | \\ H & H \end{array}\right]_n$$

monomer
chloroethene

polymer
poly(chloroethene)

Polychloro(ethene) used to be called PVC and its properties can be modified to make it a flexible covering for electrical cable insulating covering as well as pipes, etc.

quickfire

⑮ Suggest two changes that could be made to the structure of a polymer to lower its melting temperature.

Poly(phenylethene)

$$n \quad \overset{H}{\underset{H}{>}}C=C\overset{\bigcirc}{\underset{H}{<}} \longrightarrow \left[\begin{array}{cc} H & \bigcirc \\ | & | \\ C & - C \\ | & | \\ H & H \end{array}\right]_n$$

monomer
phenylethene

polymer
poly(phenylethene)

Poly(phenylethene) used to be called polystyrene and, since it is hard, it is used in many household items needing strength and rigidity. It can be made into an insulator by creating holes in the structure (expanded polystyrene).

2.6 Halogenoalkanes

Halogenoalkanes

Structure

Since halogens are more electronegative than carbon, **halogenoalkanes** are polar. The halogen is δ− and the carbon attached to the halogen is δ+ as in bromomethane

This δ+ means that the carbon is electron deficient and hence susceptible to **nucleophilic** attack. This leads to substitution.

$$H-\underset{\underset{H}{|}}{\overset{\overset{H}{|}}{C}}{}^{\delta+}-\overset{\delta-}{Br}$$

Reaction with aqueous sodium hydroxide

Aqueous sodium hydroxide provides a source of the OH⁻ ion. This has lone pairs and therefore acts as a nucleophile. The mechanism for the reaction involves the

$$C_3H_7\underset{\underset{H}{|}}{\overset{\overset{H}{|}}{C}}{}^{\delta+}-\overset{\delta-}{Cl} \longrightarrow C_3H_7\underset{\underset{H}{|}}{\overset{\overset{H}{|}}{C}}-OH + Cl^-$$

:OH⁻

donation of a lone pair on the OH⁻ to the δ+ carbon in the halogenoalkane.

The nucleophile attacks the δ+ carbon, donates a lone pair and forms a bond to the carbon. The carbon to chlorine bond breaks to give the chloride ion.

Overall this reaction is nucleophilic **substitution** with 1-chlorobutane producing butan-1-ol. The reaction can also be achieved using the OH⁻ in water and so can be classified as **hydrolysis**. Using water, the reaction is slow since the concentration of OH⁻ is very low. To achieve hydrolysis the halogenoalkane is heated under reflux with water or aqueous sodium hydroxide.

Effect of changing the halogen

If the halogen in a particular halogenoalkane is changed, the hydrolysis reaction will occur at a different rate. Two factors for chlorine, bromine and iodine have to be considered in explaining this variation:

(i) Electronegativity: Chlorine is the most electronegative so that the C–Cl bond is most polar and the carbon is the most δ+. This is the carbon that is attacked by the nucleophile.

(ii) Bond strength: Chlorine is the smallest halogen so that the C–Cl bond is the strongest. In hydrolysis this bond is broken.

These two factors act in opposite directions but, in practice, the iodocompound is hydrolysed most quickly. This means that the effect of the bond strength outweighs the effect of the charge on the carbon.

Key Terms

Halogenoalkanes are a homologous series in which one or more hydrogen atoms in an alkane have been replaced by a halogen.

Hydrolysis is a reaction with water to produce a new product.

Nucleophile is a species with a lone pair of electrons that can be donated to an electron-deficient centre.

Substitution reaction is a reaction in which one atom/group is replaced by another.

Grade boost

The hydrolysis of 1-chlorobutane is one of the mechanisms required in the specification. Examiners like to ask questions that show you understand mechanisms!

quickfire

① Draw the displayed formula of 1,2-dichlorobutane.

quickfire

② Draw the skeletal formula of 3-iodopent-2-ene.

quickfire

③ Is reacting chlorine with ethane a satisfactory method of preparing pure 1-chloroethane? Explain your answer.

Test for the presence of a halogen in an organic compound

Grade boost

Make sure you can explain why iodocompounds are hydrolysed more rapidly than chlorocompounds. Remember that there are two concepts to consider and they act in opposite directions.

Grade boost

The use of aqueous silver nitrate in the test for the presence of a halogen in an organic compound is the same as you use to test for halide ions in inorganic compounds.

Since the hydrolysis of an organic compound containing a halogen produces a halide ion, the reaction can be used to show the functional group – X (where X = Cl, Br or I).

In practice, the organic compound is heated with aqueous sodium hydroxide as above.

$$RX + NaOH(aq) \rightarrow ROH + Na^+(aq) + X^-(aq)$$

R is the alkyl group (or other organic part of the molecule).

The presence of X^- (aq) can be shown by adding $AgNO_3(aq)$ but any NaOH(aq) remaining after the hydrolysis would interfere with this test and must be removed. This is done by adding $HNO_3(aq)$.

The presence of the halogen, in the original organic compound, can then be seen in the usual test for halides.

halogen	addition of $Ag^+(aq)$	addition of $NH_3(aq)$ to precipitate formed with $Ag^+(aq)$
chlorine	white precipitate	dissolves in dilute $NH_3(aq)$
bromine	cream precipitate	dissolves in concentrated $NH_3(aq)$
iodine	yellow precipitate	does not dissolve in $NH_3(aq)$

Key Term

An **elimination reaction** = one that involves the loss of a small molecule to produce a double bond.

Elimination reactions

An elimination reaction for a halogenoalkane involves the loss of hydrogen halide and the formation of a double bond. Hydrogen halide is acidic and can be removed using an alkali. To avoid hydrolysis the alkali must be in solution in ethanol.

quickfire

④ Write down, in order, the steps needed to test for the presence of iodine in an organic compound. Include the result expected.

quickfire

⑤ Write the ionic equation for the reaction that occurs to produce the white precipitate formed if chlorine is present in an organic compound.

$$
\begin{array}{c}
\text{H H H} \\
| \quad | \quad | \\
\text{H−C−C−C−H} \\
| \quad | \quad | \\
\text{Br H H}
\end{array}
\longrightarrow
\begin{array}{c}
\text{H} \\
\diagdown \\
\text{C=C−C−H} \\
\diagup \quad \quad | \\
\text{H} \quad \quad \text{H}
\end{array}
+ \text{HBr}
$$

1-bromopropane   propene

This equation can also be written as:

$$
\begin{array}{c}
\text{H H H} \\
| \quad | \quad | \\
\text{H−C−C−C−Br} \\
| \quad | \quad | \\
\text{H H H}
\end{array}
+ \text{NaOH}
\longrightarrow
\begin{array}{c}
\text{H H} \\
| \quad | \quad \diagup \text{H} \\
\text{H−C−C=C} \\
| \quad \quad \diagdown \text{H} \\
\text{H}
\end{array}
+ \text{NaBr} + \text{H}_2\text{O}
$$

In order to eliminate hydrogen halide, the halogen must be attached to a carbon next to a carbon that has a hydrogen attached.

Example

If the halogenoalkane is unsymmetrical, more than one alkene can be formed by elimination reactions. An example is:

but-1-ene

and

but-2-ene

You do not need to know which product would dominate.

Uses of halogenoalkanes

1 As solvents

Although halogenoalkanes are polar, they do not contain the O–H or N–H needed to form hydrogen bonds with water. They are therefore insoluble in water but they are able to mix with a variety of non-polar/polar organic substances. This means they can be used as solvents in a variety of processes and that they are very effective degreasing agents. This includes their use in dry cleaning. Chlorocompounds are most commonly used as they are the cheapest.

2 As anaesthetics

Trichloromethane, $CHCl_3$, was an early form of an anaesthetic when it was called chloroform. Some halogenoalkanes are still used as anaesthetics.

3 As refrigerants

Although small halogenoalkanes are gases at room temperature the presence of permanent dipole–permanent dipole attractions means that boiling temperatures are close to room temperature. They are therefore liquids that can easily be evaporated or gases that can easily be liquefied at room temperature.

When a liquid is vaporised it uses heat and this can be taken from the surroundings. These easily vaporised liquids are therefore used in refrigerators.

quickfire

⑥ Explain the observation that, when heated with sodium hydroxide dissolved in ethanol, 2-chlorobutane produces two products but 3-chloropentane only produces one. Draw the structures of all three products.

quickfire

⑦ Why are halogenoalkanes used in products designed to remove grease from hands or machines?

quickfire

⑧ How can CFCs be liquefied so that they can be used as refrigerants?

Key Terms

CFCs are halogenoalkanes containing both chlorine and fluorine.

HFCs are halogenoalkanes containing fluorine as the only halogen.

Ozone layer is a layer around the earth containing O_3 molecules.

Grade boost

Be careful not to confuse agents causing global warming with those causing ozone depletion.

Grade boost

The C–Cl bond, rather than the C–F bond, breaks because fluorine is smaller than chlorine and therefore forms stronger covalent bonds.

quickfire

⑨ How do you recognise a propagation step in a reaction mechanism?

⑩ Write the overall equation for the propagation steps shown in the destruction of the ozone layer.

⑪ Why are HFCs being increasingly used in place of CFCs?

Grade boost

Do not try to remember the propagation steps shown in the main text. Many alternatives are possible but you should recognise that overall you are changing O_3 to O_2 whilst regenerating the chlorine radical.

Environmental effects of CFCs

The halogenoalkanes most commonly used for refrigerants and propellants in spray cans contained both chlorine and fluorine. They were chlorofluorocarbons, **CFCs**. The use of such compounds is now tightly regulated because of their toxicity and their adverse effects in the upper atmosphere. Holes in the **ozone layer** are thought to have been caused by CFCs. Such holes allow uv rays to reach the earth's surface and cause skin cancer.

CFCs affect ozone in a radical chain reaction that, in a similar mechanism to the halogenation of alkanes, occurs in three stages.

Initiation stage

Initiation occurs due to bond fission of the C–Cl bond in CFC to produce radicals. This is caused by uv radiation in the upper atmosphere. It is the C–Cl bond, rather than the C–H or C–F bond, that is broken because this is the weakest of these bonds.

Using trichlorfluoromethane as an example of a CFC.

$$CFCl_3 \rightarrow Cl^{\bullet} + CFCl_2^{\bullet}$$

Propagation stage

The process is complex and there are many possible propagation reactions. These include:

$$Cl^{\bullet} + O_3 \rightarrow ClO^{\bullet} + O_2$$
$$ClO^{\bullet} + O_3 \rightarrow Cl^{\bullet} + 2O_2$$

This is a chain reaction and so the formation of a small number of chlorine radicals can cause the decomposition of many ozone molecules.

Termination stage

Since many different radicals can be formed in the propagation stage many final products are possible.

Due to problems of ozone depletion, much work has been done to find replacements for CFCs. Suggestions include the use of hydrofluorocarbons (**HFCs**) since these do not contain C–Cl bonds and cannot give chlorine radicals.

2.7 Alcohols and carboxylic acids

Alcohols

Alcohols contain the functional group –OH. Compounds with more than one OH do exist but in this unit you only need to know about those with just one. Ethanol is the most widely used alcohol and, in everyday language, is just called alcohol.

The industrial preparation of ethanol

Ethene reacts with steam to produce ethanol.

$$CH_2=CH_2(g) + H_2O(g) \rightleftharpoons CH_3CH_2OH(g) \quad \Delta H = -45 \text{ kJ mol}^{-1}$$

The conditions used are a temperature of 300°C, a pressure of 60–70 atmospheres and a catalyst of phosphoric acid (coated onto an inert solid). The conditions used can be explained using **Le Chatelier's principle**.

Temperature

The forward reaction is exothermic and so a high yield would be favoured by a low temperature. This, however, would give a slow rate of reaction, so 300°C is a compromise temperature.

Pressure

Two moles of gaseous reactants give one mole of gaseous product and so a high yield is favoured by a high pressure. This also increases the rate of reaction but too high a pressure is expensive to maintain.

Catalyst

This increases the rate of the reaction without affecting the yield. Using these conditions gives about 5% conversion of ethene to ethanol and therefore the unreacted ethene is recycled back to the reaction chamber.

Fermentation

Sugars are converted to alcohol by **fermentation**. The reaction is catalysed by an enzyme present in yeast so that the sugar is dissolved in water, yeast is added and the mixture is left in a warm place. This is the method by which alcoholic drinks are made.

Using glucose as an example of a sugar the reaction is:

$$C_6H_{12}O_6 \rightarrow 2C_2H_5OH + 2CO_2$$

Since ethanol has a boiling temperature of 80°C it can be separated from the aqueous mixture left in the fermentation vessel by fractional distillation.

Key Terms

Alcohol is an homologous series containing –OH as the functional group.

Fermentation is an enzyme-catalysed reaction that converts sugars to ethanol.

Grade boost

Make sure you can apply Le Chatelier's principle.

quickfire

 Why is a very high temperature not used in the industrial preparation of ethanol from ethene?

Grade boost

The reaction between ethene and steam is electrophilic addition. It has the same mechanism as the other reactions of alkenes.

quickfire

 Why is fermentation carried out at a temperature of about 38°C?

quickfire

③ How can ethanol be removed from the mixture produced when sugars are fermented?

Key Term

Biofuel is a fuel that has been produced using a biological source.

Biofuels

Biofuels are made from living organisms. Bioethanol is obtained by the fermentation of sugars in plants and biodiesel from the oils and fats present in the seeds of some plants.

The use of biofuels has advantages and disadvantages.

Advantages

1 Renewable

Unlike fossil fuels the plants needed to produce biofuels can be grown each year and waste material from animals can also be used.

2 Greenhouse gases

In the same way as fossil fuels, biofuels do produce carbon dioxide when they are burned. However, the plants, from which they were formed, have taken in the carbon dioxide during photosynthesis.

$$6CO_2 + 6H_2O \rightarrow C_6H_{12}O_6$$

This means that overall the use of biofuels Is carbon neutral.

3 Economic and political security

Countries without their own source of fossil fuels are affected by changes in price and availability when these fuels are imported.

Disadvantages

1 Land use

Forests are being destroyed to create land on which to grow plants for biofuels. Land used for biofuels cannot produce food crops.

2 Use of resources

Large quantities of water and fertiliser are needed to grow biofuels. Water in many areas is in short supply and the use of fertilisers can cause water pollution.

3 Carbon neutrality?

Although the use by plants and production on burning balance, this does not take Into account the fuel needed to build the factories, transport raw materials, etc.

Grade boost

You need to be able to describe advantages and disadvantages of the use of biofuels as well as considering the similarities and differences of their use compared with using fossil fuels.

Grade boost

Be careful that you understand that biofuels do produce carbon dioxide on combustion.

quickfire

④ Why can the use of biofuels be considered as not contributing to global warming?

Dehydration of primary alcohols

Many alcohols can be dehydrated to form alkenes. The equation for butan-1-ol is shown below.

$$CH_3CH_2CH_2CH_2OH \rightarrow CH_3CH_2CH=CH_2 + H_2O$$

but-1-ene

A variety of dehydrating agents can be used but the most commonly used ones are heated aluminium oxide or concentrated sulfuric acid.

Water is removed (the H from one carbon atom and the OH from the next) and a double bond is formed.

Grade boost

The dehydration of alcohols is the reverse of the reaction of alkenes with steam.
You will need to be able to quote a suitable dehydrating agent and recognise that others are possible.

quickfire

⑤ Classify the following as 1°, 2° or 3° alcohols.

(a) Ethanol CH_3CH_2OH

(b) 3-methyl pentan-3-ol $CH_3CH_2C(CH_3)(OH)CH_2CH_3$

(c) Pentan-2-ol $CH_3CH(OH)CH_2CH_2CH_3$

(d) Pentan-1-ol $CH_2(OH)CH_2CH_2CH_2CH_3$

Classification of alcohols

Alcohols are **classified** as being primary, 1°, secondary, 2°, or tertiary, 3°, according to the bonding of the —OH in the molecule.

If the —OH is joined to a carbon that is itself joined to not more than one other carbon atom, the alcohol is primary.

If the —OH is joined to a carbon that is itself joined to two other carbon atoms, the alcohol is secondary.

If the —OH is joined to a carbon that is itself joined to three other carbon atoms, the alcohol is tertiary.

Examples of classification

(i) Methanol

$$\begin{array}{c} H \\ | \\ H-C-OH \\ | \\ H \end{array}$$

The OH is joined to a carbon attached to no other carbons and so it is a primary alcohol.

(ii) Propan-1-ol

$$\begin{array}{c} H \ \ H \ \ H \\ | \ \ \ | \ \ \ | \\ H-C-C-C-OH \\ | \ \ \ | \ \ \ | \\ H \ \ H \ \ H \end{array}$$

The OH is joined to a carbon attached to one other carbon and so it is a primary alcohol.

(iii) Propan-2-ol

$$\begin{array}{c} H \ \ H \ \ H \\ | \ \ \ | \ \ \ | \\ H-C-C-C-H \\ | \ \ \ | \ \ \ | \\ H \ \ OH \ H \end{array}$$

The OH is joined to a carbon attached to two other carbons and so it is a secondary alcohol.

quickfire

⑥ Write equations to show:
 (a) The dehydration
 (b) The complete oxidation
 of propan-1-ol.

Grade boost

If you include the oxidation number of, e.g. potassium dichromate (VI), it must be correct. If you are unsure, it might be safer to omit the number.

Grade boost

You may see $Cr_2O_7^{2-}$/H^+ in a reaction scheme. This shows the oxidising agent acidified dichromate(VI).
When you use [O] in an oxidation equation, do not forget to balance the equation.

Grade boost

Make sure you recognise that the oxidation of primary alcohols is a two-stage process, that the oxidation of secondary alcohols is a one-stage process and that tertiary alcohols are not generally oxidised.

quickfire

⑦ Why are tertiary alcohols not generally oxidised?

quickfire

⑧ What is the functional group in an aldehyde?

quickfire

⑨ What would you expect to see if acidified potassium dichromate(VI) was heated with propan-2-ol?

(iv) 2-methyl butan-2-ol

$$
\begin{array}{c}
\quad\;\; H \quad OH\;H\;\;H \\
\quad\;\; | \quad\;\; | \quad | \quad | \\
H-C-C-C-C-H \\
\quad\;\; | \quad\;\; | \quad | \quad | \\
\quad\;\; H \quad\;\; | \quad H\;\;H \\
\quad\;\; H-C-H \\
\quad\quad\quad | \\
\quad\quad\quad H
\end{array}
$$

The OH is joined to a carbon attached to three other carbons and so it is a tertiary alcohol.

Oxidation of alcohols

Many alcohols can also be oxidised. The usual oxidising agent is acidified potassium dichromate(VI) which is heated with the alcohol.

What happens depends on whether the alcohol is primary, secondary or tertiary.

In organic oxidation reactions the oxidising agent is usually shown as [O] in equations.

$$
\begin{array}{c}
\;\;\;\; H \\
\;\;\;\; | \\
R-C-OH + 2[O] \longrightarrow R-C{\overset{\displaystyle O}{\underset{\displaystyle OH}{}}} + H_2O \\
\;\;\;\; | \\
\;\;\;\; H
\end{array}
$$

1 Primary – using propan-1-ol as an example

The reaction takes place in two stages.

Stage 1

$$
H-\underset{H}{\overset{H}{C}}-\underset{H}{\overset{H}{C}}-\underset{H}{\overset{H}{C}}-O-H + [O] \rightarrow H-\underset{H}{\overset{H}{C}}-\underset{H}{\overset{H}{C}}-C{\overset{\displaystyle O}{\underset{\displaystyle H}{}}} + H_2O
$$

propan-1-ol   propanal

Two hydrogen atoms are lost – one from the alcohol OH and one from the adjacent carbon. This creates a carbon to oxygen double bond.

The, product with the functional group

$$
-C{\overset{\displaystyle O}{\underset{\displaystyle H}{}}}
$$

is an aldehyde. In this case propanal.

Stage 2

The aldehyde is oxidised further.

$$
H-\underset{H}{\overset{H}{C}}-\underset{H}{\overset{H}{C}}-C{\overset{\displaystyle O}{\underset{\displaystyle H}{}}} + [O] \rightarrow H-\underset{H}{\overset{H}{C}}-\underset{H}{\overset{H}{C}}-C{\overset{\displaystyle O}{\underset{\displaystyle OH}{}}}
$$

propanal   propanoic acid

An oxygen Is added to the aldehyde. The product with the functional group

$$-C\overset{\displaystyle O}{\underset{\displaystyle O-H}{{\Large\diagdown}}}$$

is a carboxylic acid. In this case propanoic acid.

2 Secondary – using propan-2-ol as an example

The reaction has only one stage – this corresponds to stage 1 of the oxidation of primary alcohols.

propan-2-ol

The product, with the functional group $-C{=}O$, is a ketone. In this case propanone.

3 Tertiary – using 2-methylpropan-2-ol as an example

2-methylpropan-2-ol

Since there is no hydrogen on the adjacent carbon that can be lost, no reaction takes place.

Summary of oxidation reactions

Primary alcohols $\longrightarrow$ aldehydes $\longrightarrow$ carboxylic acids

Seconday alcohols $\longrightarrow$ ketones

Tertiary alcohols are not oxidised.

Use of acidified potassium dichromate(VI)

When it behaves as an oxidising agent acidified potassium dichromate(VI) changes colour from orange to green. This can be used as a test for primary and secondary alcohols since they will give a positive test result but tertiary alcohols will not.

Key Terms

Carboxylic acid is an homologous series containing –COOH as the functional group.

Weak acid is an acid that only partially ionises in aqueous solution.

Grade boost

The reactions of carboxylic acids as acids are similar to strong acids – they lose the H^+ and generally replace it with a metal to form a salt.

Grade boost

Be careful with the formula of the salt if the metal forms 2^+ ions.

quickfire

⑩ What is the formula of:
(a) Sodium ethanoate?
(b) Zinc propanoate?

Grade boost

In esterification, water is removed. To find the formula of the ester draw the alcohol and carboxylic acid with the 2 OH groups together. Draw a box round the water and join the organic section together. This is the ester.

quickfire

⑪ Draw the ester that is formed when butan-2-ol reacts with ethanoic acid.

quickfire

⑫ Complete the following: Carboxylic acids are .... because they are .... ionised. They react with alkalis to form ....

Carboxylic acids

Carboxylic acids contain the functional group

$$-C\overset{\displaystyle O}{\underset{\displaystyle O-H}{\big\backslash}}$$

They are **weak acids** and so are only partly dissociated to produce H^+ ions.

$$CH_3COOH \rightleftharpoons CH_3COO^- + H^+$$

Reactions

As acids

Carboxylic acids behave in the same way as strong inorganic acids when they react with bases, carbonates and hydrogencarbonates to form salts.

Examples

1 Bases and alkalis

General equation: Acid + Base → Salt + Water

Alkali: $CH_3CH_2COOH(aq) + KOH(aq) \rightarrow CH_3CH_2COOK(aq) + H_2O(l)$
     propanoic acid   potassium propanoate

Base: $2CH_3COOH(aq) + MgO(s) \rightarrow (CH_3COO)_2Mg(aq) + H_2O(l)$
     ethanoic acid   magnesium ethanoate

2 Carbonates and hydrogencarbonates

Both behave in a similar way.

General equation: Acid + Carbonates → Salt + Carbon dioxide + Water

Carbonate: $2HCOOH(aq) + CuCO_3(s) \rightarrow (HCOO)_2Cu(aq) + CO_2(g) + H_2O(l)$
        methanoic acid   copper(II) methanoate

Hydrogencarbonate:

$$CH_3COOH(aq) + KHCO_3(aq) \rightarrow CH_3COOK(aq) + CO_2(g) + H_2O(l)$$
                       potassium ethanoate

Esterification

Carboxylic acids also react to form esters.

General equation: Acid + Alcohol $\rightleftharpoons$ Ester + Water

The reaction is catalysed by concentrated sulfuric acid so that the alcohol, carboxylic acid and concentrated sulfuric acid are heated together. The ester can be separated by distillation.

$$CH_3COOH + C_2H_5OH \rightleftharpoons CH_3COOC_2H_5 + H_2O$$
   ethanoic acid   ethanol   ethyl ethanoate

Esters are recognised since they have characteristic sweet, fruity odours.

2.8 Instrumental analysis

Traditionally, to analyse the nature and quantity of an unknown substance, volumetric and gravimetric analysis techniques were used. In volumetric analysis, titrations are used to measure volumes of solutions, whilst in gravimetric analysis, masses of solids are found. Nowadays spectroscopic techniques are more often the preferred method.

In this unit you need to be able to analyse mass, IR and NMR spectra to help identify the structure of an organic molecule.

Mass spectrometry

In a mass spectrometer an electron is knocked off molecules of an organic compound to produce a positive ion. This is the **molecular ion** and is often shown as M^+. The mass spectrometer also causes the molecules to split into smaller parts – **fragments**. These fragments can give information about the structure of the molecule.

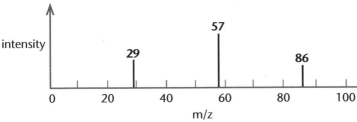

The x axis in a mass spectrum shows mass/charge (m/z) but you can assume that the charge on the ion is 1. To interpret a mass spectrum, look at the peak with the largest m/z. This is the molecular ion and it gives the M_r.

The spectrum above is that of pentan-3-one, $C_2H_5C=OC_2H_5$. This has a M_r of 86 and this corresponds to the peak with the largest m/z. The peak at 29 is due to $C_2H_5^+$ and, if this is knocked off the molecule, a fragment of mass 57 remains. This is due to $C_2H_5C=O^+$.

If chlorine or bromine are present in the compound, two peaks for M^+ and some of the fragments will be seen, since both halogens can exist as two isotopes.

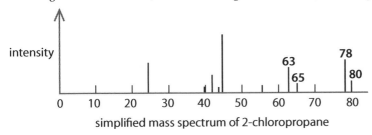

simplified mass spectrum of 2-chloropropane

From the spectrum, M^+ peak is at 78 for $C_3H_7{}^{35}Cl$ but at 80 for $C_3H_7{}^{37}Cl$. The peaks at 63 and 65 are due to the loss of 15 i.e. CH_3^+ from the molecule.

The other fragment peaks are caused by rearrangements and are difficult to interpret from the structure of 2-chloropropane.

Key Terms

Fragmentation is splitting of molecules in a mass spectrometer into smaller parts.

Molecular ion is the positive ion formed in a mass spectrometer from the whole molecule.

quickfire

① Why are spectroscopic techniques generally preferred to volumetric and gravimetric methods nowadays?

quickfire

② What information is given by the y-axis values in a mass spectrum? Is this useful in finding the structure of the compound?

Grade boost

Subtraction of m/z values can show what has been broken off – for example, in the pentan-3-one mass spectrum 86 − 57 = 29 shows the loss of C_2H_5.

Grade boost

The molecular ion has the largest m/z. Do not say it is the largest peak.

quickfire

③ At what m/z value would you expect to see the largest m/z value peak for ethanol?

Grade boost

Rearrangements of parts of the molecule can occur. Do not try to explain the identity of all the peaks in a mass spectrum in an exam. You will be asked about those that help give the structure.

Key Terms

Wavenumber is a measure of energy absorbed, used in IR spectra.

Characteristic absorption is the wavenumber range at which a particular bond absorbs radiation.

Grade boost

It is the troughs, i.e. peaks downwards, which are used in IR spectra.

quickfire

④ (a) Why are there peaks at m/z 78 and 80 in the IR spectrum of 1-chloropropane?

(b) Why are the peaks at 78 and 80 not the same height in the mass spectrum of 1-chloropropane?

quickfire

⑤ At what m/z would you expect to see the largest value for butan-1-ol?

What group has been lost from butan-1-ol to give a peak with m/z at 59?

quickfire

⑥ Why is an absorption in the range 2800 to 3100 not really useful in determining the groups present?

quickfire

⑦ In which of the compounds

$$CH_3C \overset{O}{\overset{\|}{-}} OCH_3$$

and CH_3CH_2OH would you expect to see an absorption at 1700 to 1720 cm^{-1}?

Infrared spectroscopy

Radiation in the IR part of the spectrum is absorbed to cause increased vibrations and bending in organic molecules.

The **wavenumber** at which the **absorption** occurs is **characteristic** of the bond and therefore useful in identifying the functional group present.

You will be given the wavenumbers you need to answer exam questions. Examples of these are shown in the table.

Bond	Wavenumber /cm^{-1}
C–C	650 to 800
C–O	1000 to 1300
C=O	1650 to 1750
C–H	2800 to 3100
O–H	2500 to 3550

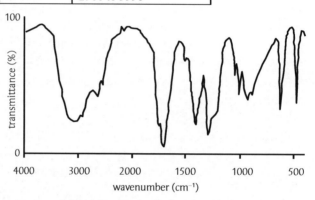

infra-red spectrum of ethanoic acid, CH_3COOH

In the same way that not all peaks were useful in mass spectra, not all absorptions are useful in interpreting IR spectra. Just look at the information you have and look for absorptions corresponding to the functional groups/ bonds present.

The spectrum of ethanoic acid shown above is consistent with the structure of ethanoic acid since it shows an absorption at approximately 1750 cm^{-1} from C=O, one at approximately 1250 cm^{-1} from C–O and one at approximately 3000 cm^{-1} from O–H.

It would have been difficult, without some other Information, to positively identify ethanoic acid just from the above spectrum using individual peaks. However, even complex molecules can be identified using a database. These show the whole spectrum of a huge range of organic molecules so that unknowns can be matched to these.

Nuclear magnetic resonance spectroscopy

Energy is absorbed to change the spin of atoms but you do not need to know exactly what Is happening to the atom. The important fact is that the energy absorbed depends on the **environment** in which the atom is. Each absorption will appear at a different place on the spectrum and the δ value (the **chemical shift**) for that peak tells you the energy involved. There are two types of nuclear magnetic resonance spectra in this unit ^{13}C and ^{1}H.

Key Terms

Chemical shift is a measure of difference, in parts/million, from the standard of the energy of a particular absorption type.

Environment is the nature of the surrounding atoms/groups in a molecule.

^{13}C Spectroscopy

The presence of very small amounts of ^{13}C in organic compounds means that they will absorb energy and produce a spectrum.

As with infrared spectra you will be given the data you need to interpret the spectrum. Some δ values are given in the table.

Type of carbon	Chemical shift, δ/ppm
C—C	5 to 55
C—O	50 to 70
C—Cl	30 to 70
C=C	115 to 145
C=O	190 to 220

An example of a ^{13}C spectrum is that for but-3-en-2-one,

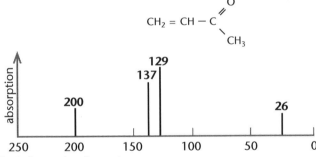

The information from the spectrum:

- Number of peaks gives the *number* of different carbon environments.
- The chemical shift of the peaks gives the *type* of carbon environment.

From the spectrum there are four peaks and therefore C atoms in four environments.

Looking at the chemical shifts to find the environments:

peak at δ = 200 ppm due to **C**=O

peaks at δ = 137 ppm and 129 ppm due to **C** =C and C=**C**

peak at δ = 26 ppm due to **C**H_3

It is not possible, using the information available, to tell which of the C=C peaks is at which end of the double bond.

Grade boost

In both ^{13}C and ^{1}H spectra count the number of peaks – this tells you how many different environments there are in the molecule.

Grade boost

δ values can be used to identify the type of environment but, because these are influenced by the presence of other groups, they are often found significantly away from what the data table suggests.

quickfire

(8) What can you work out from the peak heights in:
(a) a ^{13}C NMR spectrum?
(b) a ^{1}H NMR spectrum?

¹H Spectroscopy

This is sometimes called proton spectroscopy and, as for ^{13}C, it gives Information about:

- The *number* of different proton environments – from the number of peaks.
- The *types* of proton environments - from the chemical shifts.

However, ¹H spectra also give information about the ratio of the numbers of protons in each environment.

Some δ values are given in the table.

Type of proton	Chemical shift, δ/ppm
$R{-}CH_2$	0.7 to 1.6
$R{-}OH$	1.0 to 5.5
$R{-}CH_2{-}R$	1.2 to 1.4
$R_3{-}CH$	1.6 to 2.0
$-C{=}OCH-$	1.9 to 2.9
$-O-CH_3$ $-O-CH_2R$ $-O-CHR_2$	3.3 to 4.3
$-CHO$	9.1 to 10.1

An example of a ¹H spectrum is that for methyl propanoate, $CH_3CH_2COOCH_3$.

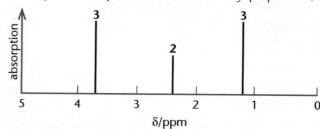

From the spectrum there are three peaks and therefore H atoms in three environments.

The H atoms in each environment are in the ratio 3:2:3.

Looking at the chemical shifts to find the environments:

peak at δ = 1.2 ppm due to $\mathbf{CH_3}CH_2C{=}O$

peak at δ = 2.4 ppm due to $CH_3\mathbf{CH_2}C{=}O$

peak at δ = 3.7 ppm due to $O{-}\mathbf{CH_3}$

Summary: Energy, Rate and Chemistry of Carbon Compounds

2.1 Thermochemistry

Enthalpy changes

- Enthalpy change, ΔH, is the heat added to a system at constant pressure
- If ΔH is negative the reaction is exothermic. If ΔH is positive the reaction is endothermic
- Important enthalpy changes are: formation, $\Delta_f H$, combustion, $\Delta_c H$, reaction, $\Delta_r H$
- Hess's law states that the total enthalpy change for a reaction is independent of the route taken from the reactants to the products. It is used to calculate enthalpy changes indirectly
- Average bond enthalpy is the average value of the enthalpy required to break a given type of covalent bond in any molecule
- Calculations of enthalpy changes of reaction involving covalent compounds using average bond enthalpies will not be as accurate as results derived from experiments with specific molecules

Measuring ΔH experimentally

- Use $q = mc\Delta T$ and $\Delta H = -q/n$ where ΔT, the corrected temperature change, can be found graphically by extrapolating back to the mixing time
  m is the mass of the solution not the solid
  n is the number of moles reacting i.e. it is the number of moles of reactant not in excess
- To prevent heat from escaping or being absorbed, a polystyrene cup with a lid is used instead of a beaker

2.2 Rates of reaction

- Rate of reaction is the change in concentration of a reactant or product per unit time
- It can be calculated by plotting a graph and finding the gradient
- Collision theory states that for a chemical reaction to take place, molecules must collide successfully, i.e. with a minimum amount of energy known as the activation energy
- Activation energy may be shown on energy profile diagrams
- An increase in concentration of a solution, pressure of a gas or surface area of a solid increases the rate of reaction because there is a greater chance of successful collisions in a certain length of time
- An increase in temperature increases the rate of reaction because more colliding molecules have the required activation energy. This can be shown using energy distribution curves

Catalysts

- A catalyst increases the rate of reaction by providing an alternative route of lower activation energy
- A heterogeneous catalyst is in a different phase from the reactants. A homogenous catalyst is in the same phase as the reactants
- Enzymes are being increasingly used in industrial processes as they achieve some of the goals of Green Chemistry

Measuring reaction rates experimentally

- Rates can be followed experimentally by a change in gas volume or pressure, a change in mass or a change in colour
- Gas collection is a good way of showing how rate changes during a chemical reaction
- 'Iodine clock' and precipitation reactions are good experiments to compare rates of reaction under different conditions in order to establish the relationship between reactant concentrations and rate

2.3 The wider impact of chemistry

Impacts

- Social: Location of workers and industries
- Economic: Cost, requirement and employment
- Environmental: Local and global pollution
- Energy production: Need and problems

Energy problems

- Fossil fuels, carbon dioxide generation and climate change
- Carbon neutrality, renewable biofuels
- Nuclear and solar power
- The hydrogen economy

Green Chemistry

- Renewable raw materials as from plants
- Enzyme catalysts to save energy at low temperatures
- High atom economy
- Avoid toxic chemicals and solvents

General comment

- This unit is not about memorising facts but the discussion and analysis of given situations and seeing how science can work to deal with them

2.4 Organic compounds

Types of formulae

- Molecular formula shows the type and number of the atoms in the molecule
- Displayed formula shows all the atoms and bonds in the molecule
- Shortened formula shows the groups in a molecule in sufficient detail so that the structure is unambiguous
- Skeletal formula shows the carbon/hydrogen backbone of the molecule and any functional groups attached

Nomenclature rules

- The name of an organic compound is based on the number of carbon atoms in the longest chain and the functional group present
- The number of carbons in the longest chain is shown by use of meth, eth, prop, but, pent, hex, hept, oct, non and dec
- The functional group is shown by ene (C=C), ol (OH), halogeno (any halogen), oic acid (COOH)

Physical properties

- Physical properties are affected by chain length and the presence of functional groups
- Melting and boiling temperatures increase with increasing M_r within an homologous series

Structural isomerism

- Structural isomers are compounds with the same molecular formula but different structural formulae

Mechanisms

- The mechanism of a reaction can be classified according to the nature of the attack and the overall change in the nature of the organic reagent

2.5 Hydrocarbons

Alkanes

- Alkanes are used as fuels since ΔH of combustion is negative but there are disadvantages in this use of fossil fuels
- All bonds are σ bonds
- Alkanes are generally unreactive but will undergo photochlorination. This involves a radical mechanism leading to substitution

Alkenes

- The carbon to carbon double bond in an alkene consists of a σ and a π bond
- The π bond is a region of high electron density and this is susceptible to electrophilic attack leading to addition across the double bond
- E-Z isomers of many alkenes exist because there is limited rotation about the π bond
- The presence of the double bond can be tested for by the decoloration of bromine or of potassium manganate(VII)
- The product of the reaction between HBr and an unsymmetrical alkene can be predicted by considering the fact that secondary carbocations are more stable than primary carbocations
- Alkenes can be hydrogenated to reduce or remove unsaturation
- Many alkenes and substituted alkenes undergo addition polymerisation to give a wide range of commercially important polymers

2.6 Halogenoalkanes

Reactions

- Sodium hydroxide, dissolved in ethanol, can be used to eliminate hydrogen halide from a halogenoalkane – this gives an alkene
- Halogenoalkanes are susceptible to nucleophilic attack on the $\delta+$ carbon – this leads to substitution of the halogen. An example of this is the use of aqueous sodium hydroxide to give an alcohol
- The rate of nucleophilic substitution reactions depends on the nature of the halogen. The iodocompound reacts faster than the corresponding compounds of the other halogens due to the relative weakness of the C to I bond
- When halogenoalkanes are hydrolysed the halide ion is formed. The halogen present can then be identified using the test with aqueous silver nitrate

Uses of halogenoalkanes

- Halogenoalkanes were widely used as solvents, anaesthetics and refrigerants because of their properties but nowadays their use is tightly regulated
- The relative adverse environmental effects of different CFCs can be related to the strength of the bonds present

Practical work

- The hydrolysis of 1-bromobutane to produce butan-1-ol can be achieved by refluxing the halogenoalkane with aqueous sodium hydroxide

2.7 Alcohols and carboxylic acids

Preparation of ethanol and biofuels

- Ethanol can be produced by hydration of ethene or by fermentation of sugars. Knowledge of the general conditions used in each case is required

- Bioethanol and biodiesel can both be used as alternatives to fossil fuels but there are both advantages and disadvantages in doing so

Classification and reactions of alcohols

- The functional group in an alcohol is –OH

- Alcohols can be classified as primary, secondary or tertiary according to what else is attached to the carbon to which the alcohol OH is attached

- Most alcohols can be dehydrated to produce alkenes – a variety of dehydrating agents can be used

- Primary and secondary alcohols can be oxidised by oxidising agents such as acidified potassium dichromate(VI). Primary alcohols are oxidised to aldehydes and then to carboxylic acids whilst secondary alcohols are oxidised to ketones

Reactions of carboxylic acids

- Carboxylic acids are weak acids with –COOH as the functional group

- Carboxylic acids react with bases to produce salt and water. They also react with carbonates and hydrogencarbonates to produce salt, carbon dioxide and water

- Carboxylic acids react with alcohols to produce esters – this reaction is catalysed by concentrated sulfuric acid

Practical work

- Esters can be made by heating, generally under reflux, a carboxylic acid, an alcohol and concentrated sulfuric acid. This reaction gives an equilibrium mixture. The ester can be separated from this reaction mixture by distillation

2.8 Instrumental analysis

Mass spectra

- A mass spectrum gives the M_r of the compound present by the value of m/z for the peak with the highest m/z

- Peaks at other m/z values give the masses of fragments and hence clues about the structure of the compound

IR Spectra

- The absorption peaks in IR spectra arise from energy changes in the bonds of the molecule

- The size of the energy absorption, in cm^{-1}, gives information about the functional group present

NMR spectra

- Energy changes, in ppm, can be seen on a spectrum using ^{13}C or ^1H present in compounds

- The number of peaks in a ^{13}C spectrum gives the number of carbon environments whilst the chemical shift of each peak gives the type of environment

- The number of peaks in a ^1H spectrum gives the number of hydrogen environments whilst the chemical shift of each peak gives the type of environment. The ratio of the peak heights gives the ratio of the number of hydrogen atoms in each environment

Exam Practice and Technique

Aims

To encourage students to

- develop their interest and enthusiasm for Chemistry and in further study and careers
- develop essential knowledge and understanding of different parts of Chemistry and how they relate
- develop and demonstrate an understanding of How Science Works
- appreciate how society makes decisions about scientific issues and how science contributes to the success of the economy and of society.

AS Chemistry – a summary of assessment

Assessment here comprises two written papers of 90 minutes each, one for each of the two equal components that are both worth 50% and have 80 marks available on each paper.

Component 1 covers the Language of Chemistry, the Structure of Matter and Simple Reactions.

Component 2 deals with Energy, Rate and the Chemistry of Carbon Compounds.

Each paper consists of a Section A with short answer questions for 10 marks and a Section B having structured and extended answer questions for 70 marks.

There are no multiple-choice questions in these papers.

Assessment objectives (AOs) and weightings

Examination questions are written to reflect the AOs described in the specification and you must meet the following AOs in the context of the subject content which is given in detail in the specification.

AO1 Covers showing knowledge and understanding of all aspects of the subject.

Candidates should recognise and understand chemical knowledge and select, organise and communicate relevant information.

AO2 Covers applying this knowledge and understanding theoretically, practically, qualitatively and quantitatively.

Candidates should analyse and evaluate chemical knowledge and processes, apply these to unfamiliar situations and assess their validity and reliability.

AO3 Covers analysis, interpretation and evaluation of scientific information and evidence, making judgements, reaching conclusions and developing practical design and procedures safely, accurately and ethically. **Candidates should be aware of How Science Works and reliably record and communicate observations and measurements.**

The approximate weighting of these objectives – which is the same for both components – is as follows:

AO1 – 19%; AO2 – 21%; AO3 – 11%, giving an overall weighting for both papers of AO1 – 37%; AO2 – 42%; AO3 – 21%.

Note that a maximum of 10% will rely on recall only without any understanding.

Mathematical skills

These will be assessed throughout both papers and have a total weighting of at least 20%. They include algebraic manipulation, use of calculators, means and significant figures, graph plotting and analysis, spectral analysis and understanding of 2d and 3d structures in molecular shapes.

The requirement will be equivalent to Level 2 or GCSE.

Practical work

This essential part of chemistry will be assessed in two ways: first as a part of the written papers with a weighting of at least 15% and, secondly, through direct laboratory work that, although not awarded a mark, must be performed satisfactorily, written up in a file and assessed by your teacher.

The type of laboratory work to be done is listed in the specification and will be selected by your teacher. It will include preparations, titrations and measuring enthalpy changes and rates of reaction. The focus will be on the use and application of the scientific method and practices, data analysis and the use of instruments and equipment.

Eduqas website

This site, at www.eduqas.co.uk, contains much useful information, including practical guidance sheets for the 24 specified practicals from which your laboratory work will be selected, the specification and specimen assessment material, past papers and mark schemes and examiners' reports.

Exam tips

Questions are worded very carefully so that they are clear, concise, unambiguous and on the specification. However, candidates tend to penalise themselves unnecessarily by misreading questions either too quickly or too superficially. It is essential that the precise meaning of each word is understood; also, the mark value at the end of each question part provides a useful guide as to how much information is required.

1 **Read the questions carefully**. It is easy to misinterpret a question under exam pressure so read every word carefully and highlight key words if this helps you to focus.

2 **Look at the mark allocation** (see above). If a question part is worth three marks then you must give three points to gain these.

3 **Understand the information.** Not more than 10% of the marks at AS are obtainable through recall of information only. In addition, you may encounter unfamiliar material, but do not panic, rather think carefully and take your time to apply the principles you have learned to answer this type of question.

4 **Understand the instructions.** You must be clear about the exact meaning of the command words in the questions. Some, such as

calculate describe state explain predict complete

will be fairly obvious in the context of the question but some, such as

Name are more specific. 'Name a compound' for example requires a one-word answer of the name, not a formula and vice versa if a formula is asked for. Do not repeat the question or make a sentence, just give one word.

Compare If asked for a comparison of two things do not just describe them separately, they must be directly compared in unified phrases.

Suggest Give your thoughts and opinion, there may not be a definite answer.

In all cases do not be too brief. If you have more relevant knowledge and information to give, give it.

5 **Structured questions.** These are in several parts, usually on a common theme and increasing in difficulty through the question.

6 Special six-mark questions

Each question paper will include a special six-mark section in a question. This will deal with any or all of the AOs and be the largest single mark in the given question. A summary of the general type of mark scheme laid down for this section is as follows;

5–6 marks

All main features are described and explained and evidence provided.

3–4 marks

Main features are described and a simple explanation of how they arise given.

1–2 marks

Only main features are described or a simple explanation of how they arise given.

0 marks

No attempt made or the answer not worthy of credit.

Maximising your performance

- Know the Specification
- Work on past papers
- Do plenty of calculations
- Give yourself enough time to absorb and understand the topic. Last-minute swotting is useless; your subconscious mind will work for you if it is given time.
- Read the questions slowly and carefully; do not miss a part in your haste.
- Don't leave sections blank, your guess may be worth a mark and there are no minus marks.
- Give yourself enough time to check back over your answers, especially calculations.
- Are your numerical answers sensible and realistic?
- The material in one part of a question may give a clue as to the answer in another part, look up and down the paper.

Where students go wrong

- Assume that the examiner is an idiot; spell everything out, leaving no uncertainty.
- Never say 'it' use the name.
- Leave no doubt in the examiner's mind, e.g., in 'describe a test to distinguish between chloride and bromide ions in dilute nitric acid' the answer has four parts'
  1 the test to use,
  2 the result for chloride,
  3 the result for bromide,
  4 compare the results of 2 and 3.

  Students sometimes say that the chloride precipitate dissolves in dilute ammonia but don't mention that the bromide does not.
- Words in bold require exact answers, e.g. **name** means name only, **three significant figures** means that anything else will be wrong, e.g., 85.5 correct, 85.55 wrong, 85 wrong.
- Questions may say 'what is observed?' The answer must be what you would see, e.g., 'bubbles' and not 'CO_2 is produced' or 'a white precipitate is formed' not '$AgCl$ is produced'.

Finally

Athletes and musicians who want to excel have to work and practise hard; it is the same for you. Sitting in a group in the library, chatting and idly turning over your notes is not studying but passing the time of day. Work is hard concentration and plenty of repetition until you have the topic mastered.

To do well you need a good set of notes to work on, use the Revision Checklist columns, keep going back to check and testing your knowledge on past questions and calculations. None of us is keen on work but through work comes mastery, understanding and enjoyment in the confidence that you will get good grades.

Good luck!

Questions and answers

This part of the guide looks at student answers to examination-style questions through the eyes of an examiner. There is a selection of questions on topics in the AS specification with two sample answers – one of a high grade standard and one of a lower grade standard in each case. The examiner commentary is designed to show you how marks are gained and lost so that you understand what is required in your answers.

Radiation

1 In April 1986 there was an accident at the nuclear plant in Chernobyl. The table below shows the names, symbols, half-lives and type of emission of some of the radioactive isotopes that were found in the clouds and rain over the Welsh hills soon after the accident.

Element	Symbol of isotope	Half-life of isotope	Type of emission
caesium	^{137}Cs	30 years	β and γ
strontium	^{90}Sr		β
iodine	^{131}I	8 days	β
ruthenium	^{106}Ru	374 days	β

(a) State how a magnetic field affects radiation. [2]

(b) State what happens in the nucleus of an atom when a β-particle is emitted. [1]

(c) Another isotope of ruthenium, ^{95}Ru, decays by electron capture. Give the mass number and symbol of the product of the radioactive decay of ^{95}Ru. [1]

(d) Calculate the half-life of ^{90}Sr if it takes 84 years for 2.0 g of ^{90}Sr to be reduced to 0.25 g of ^{90}Sr. [1]

(e) After studying the table, Edmund stated 'in 2015 we should still be concerned about radioactive contamination from Chernobyl in some parts of the Welsh hills'. Is he correct? Justify your answer. [3]

Rhiannon's answer

(a) α and β-particles are deflected by a magnetic field, γ-rays are unaffected. ✓✗ ①

(b) A neutron decays. ✗ ②

(c) ^{95}Tc ✓

(d) 2 g to 0.25 g decreases by factor of 8, so half-life is 84/8 = 10.5 years ✗ ③

(e) Edmund is correct because radiation emitted by decaying caesium could cause cancer. ✓✗ ④ Caesium's half-life of 30 years means that the danger may exist for a considerable time. ✓

Examiner commentary

① Rhiannon needs to state that α and β-particles are deflected by a magnetic field in opposite directions.

② Rhiannon has not been sufficiently specific. She needs to state that a neutron decays to form a proton and an electron.

③ Dividing the initial mass by the final mass does not give the number of half-lives.

④ Rhiannon needs to state how the isotope could enter the body to get full marks.

Rhiannon achieves 4 out of 8 marks.

David's answer

(a) α-particles and β-particles are deflected by a magnetic field but in opposite directions while γ-rays are unaffected. ✓✓

(b) A nucleus loses an electron. ✓ ①

(c) Mass number 95, symbol Rh ✗ ②

(d) $2g \rightarrow 1g \rightarrow 0.5g \rightarrow 0.25g$ is 3 half-lives. So half-life is 84/3 = 28 years. ✓

(e) Edmund is correct since caesium's half-life is 30 years any radiation emitted that entered the soil and contaminated the grass would last for a long time. ✓ If people ate sheep that ate the grass they could be contaminated. ✓✓

Examiner commentary

① This is acceptable but a better answer would be a neutron decaying to a proton releasing an electron.

② This is what forms with β-emission.

David achieves 7 out of 8 marks.

Ionisation energies

Q & A

2 The graph below shows the logarithm of the successive ionisation energies of oxygen plotted against the number of the electron removed.

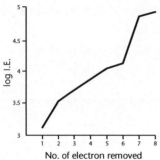

No. of electron removed

(a) Explain the shape of the graph in terms of the electronic structure of oxygen. [3]

(b) Write an equation to represent the second ionisation energy of oxygen. [1]

(c) Explain why oxygen's first ionisation energy is:

    (i) higher than that for carbon [2]

    (ii) lower than that for nitrogen [2]

    (iii) higher than that for sulfur. [2]

Rhiannon's answer

(a) A big jump in ionisation energy means that an electron has been removed from a new shell. ✓ There is a big jump after removing 6 electrons, so there are 6 electrons in the outer shell and 2 electrons in the inner shell. ✓✗ ①

(b) $O^-(g) + e \rightarrow O^{2-}(g)$ ✗

(c) (i) Oxygen has more protons so nucleus pulls electrons tighter. ✓✗ ②

    (ii) Repulsion between paired electrons in oxygen makes it easier to remove one of the electrons. ✓✗ ③

    (iii) Oxygen has one shell less so outer electron is closer and there is less shielding. ✓✓

Examiner commentary

① To gain the third mark, Rhiannon needs to mention the different orbitals in the outer shell.

② Rhiannon needs to state that the outer electrons are in the same shell or that there is little extra shielding.

③ Rhiannon has not made it clear that there are only unpaired electrons in nitrogen's outer orbital.

Rhiannon achieves 6 out of 10 marks.

David's answer

(a) There is a large jump after removing 6 electrons, so there are 6 electrons in the outer shell and 2 electrons in the inner shell. ✓✓ There is not much difference between electron 5 and 6 so these two electrons are in the same orbital in the outer shell. ✓

(b) $O^+(g) + e \rightarrow O^{2+}(g)$ ✓

(c) (i) Oxygen has a greater nuclear charge and there is no extra shielding. ✓✓

    (ii) Nitrogen only has unpaired electrons while oxygen has a pair of electrons in the 2p orbital. ✓ The repulsion between paired electrons makes it easier to remove one of the electrons. ✓

    (iii) Because its electron is closer to the nucleus and there is less shielding. ✓✗ ①

Examiner commentary

① David needs to specify that the **outer** electron is closer to the nucleus to gain both marks.

David achieves 9 out of 10 marks.

Atomic hydrogen spectrum

3 (a) Describe the visible emission spectrum of atomic hydrogen and explain how its features relate to electronic levels within the hydrogen atom. *[5]*

(b) Explain how the ionisation energy of hydrogen can be derived from the Lyman series in the atomic hydrogen spectrum. *[3]*

(c) Given that the molar ionisation energy of hydrogen is 1310 kJ mol^{-1}, calculate the value of the frequency at the start of the continuum in the hydrogen emission spectrum. *[3]*

Rhiannon's answer

(a) The spectrum is a pattern of separate lines. ✓✗ ①
Because the lines are separate it shows that the energy levels in the atom are quantised. ✓✗✗ ②

(b) The energy lines converge to a limit as they go from $n = 1$ to $n = \infty$ and using $\Delta E = hf$ this represents the ionisation energy. ✓ ③

(c) $\Delta E = hf$ therefore $f = \dfrac{\Delta E}{h} = \dfrac{1310}{6.63 \times 10^{-34}} = 1.98 \times 10^{36}$ Hz ✓✗✗ ④

Examiner commentary

① Rhiannon gains one mark for the description. To gain the second mark she needs to describe the pattern that the lines form.

② Rhiannon gets one mark for correctly using 'quantised'; however, she has not explained how the lines form or why they get closer.

③ Rhiannon's answer is very vague. Although she has stated the relationship between ionisation energy and the Lyman series, she has not explained how the ionisation energy can be derived.

④ Rhiannon has not realised that molar ionisation energy = ionisation of an atom × Avogadro's constant and has not changed I.E. into J mol^{-1}.

Rhiannon achieves 4 out of 11 marks.

David's answer

(a) The visible emission spectrum of atomic hydrogen is a series of lines ✓ which become closer as the energy increases. ✓ The lines are caused by excited electrons dropping back to a lower energy level ✗ ① therefore electrons exist in discrete energy levels. ✓ As energy increases, the energy levels become closer so the lines become closer. ✓

(b) For the Lyman series, $n = 1$, the convergence limit represents the ionisation of the hydrogen atom. ✓ The convergent frequency (difference from $n = 1$ to $n = \infty$) can be measured ✓ and the ionisation energy can be calculated from $\Delta E = hf$. ✓

(c) I.E. $= L\,\Delta E$ and $\Delta E = hf$ therefore I.E. $= Lhf$
$f = \dfrac{\text{I.E.}}{Lh} = \dfrac{1\,310\,000}{(6.02 \times 10^{23})(6.63 \times 10^{-34})} = 3.28 \times 10^{15}\ \text{s}^{-1}$ ✓✓✓ ②

Examiner commentary

① However, in the visible emission spectrum (Balmer series) the lines result from electronic transitions from higher energy levels to energy level $n = 2$.

② David has correctly converted I.E. into J mol^{-1} since the unit for Planck's constant is Js.

David achieves 10 out of 11 marks.

Mass spectrometer

4 The mass spectrometer is an important analytical instrument.

(a) Explain how a mass spectrometer works. [6]

(b) Lithium has two naturally occurring isotopes, ^{6}Li and ^{7}Li. Explain how the mass spectrum of lithium can be used to calculate the relative atomic mass of the element. [2]

Rhiannon's answer

(a) The sample is hit by electrons from an electron gun to form ions. ✓ The ions pass through a slit and are deflected by a magnetic field, ✓ the lighter the ion the greater the deflection. Changing the magnetic field brings ions of different mass/charge value to the detector in turn, where the signal is amplified and recorded. ✓✓ ①

(b) The mass spectrum will show the number of isotopes of lithium and the relative abundance of each isotope. These can be used to calculate the relative atomic mass. ✓✗ ②

David's answer

(a) First the sample is vaporised ✓ and passes into an ionisation chamber where it is bombarded by electrons to form positive ions. ✓✓ The ions are accelerated, then deflected by a magnetic field. The amount of deflection depends on the mass/charge ratio. ✓ The smaller the mass/charge ratio the greater the deflection. The beam of ions is detected electronically where their number which is proportional to the ion current is measured. ✓ The whole process takes place under high vacuum to stop air particles from colliding with sample particles. ✓ ①

(b) The mass spectrum of lithium will show the mass of each isotope and their relative percentage abundance. ✓✓

Relative atomic mass =
$$\frac{(6 \times \text{relative abundance}) + (7 \times \text{relative abundance})}{100}$$

Examiner commentary

① Although Rhiannon's explanation is correct it is not detailed enough to gain more marks. She needs to state that the sample must be gaseous when it enters the mass spectrometer, positive ions form, the ions are accelerated before being deflected and the mass spectrometer operates under a high vacuum.

② Rhiannon needs to show how this information is used to gain the second mark.

Rhiannon achieves 5 out of 8 marks.

Examiner commentary

① David has an excellent grasp of this topic and has given a full answer. Although David has not stated how the ions are accelerated he has done enough to gain full marks.

David achieves 8 out of 8 marks.

In a 6-mark question, there will be more marking points than marks available in an indicative content and the marks will be banded into three groups. To achieve the top band (5–6 marks) a candidate needs to construct a relevant, coherent and logically structured account including all key elements of the indicative content. A sustained and substantiated line of reasoning is evident and scientific conventions and vocabulary are used accurately throughout.

David clearly achieved this and scored full marks.

Rhiannon's answer contained most of the key elements and some reasoning was evident in the linking of key points therefore she falls in the next band (3–4 marks).

Moles

5 Baking powder contains mainly sodium hydrogencarbonate.

4.75 g of baking powder are contained in 250 cm³ of solution. A 25.0 cm³ portion of the aqueous solution required 29.9 cm³ of 0.170 mol dm⁻³ aqueous hydrochloric acid solution for neutralisation.

Sodium hydrogencarbonate reacts with aqueous hydrochloric acid according to the equation:

$$NaHCO_3 + HCl \longrightarrow NaCl + H_2O + CO_2$$

(a) Calculate the percentage by mass of sodium hydrogencarbonate in the baking powder. [5]

(b) Hydrogen sulfide reacts with bismuth nitrate according to the equation:

$$3H_2S(g) + 2Bi(NO_3)_3(aq) \longrightarrow Bi_2S_3(s) + 6HNO_3(aq)$$

(i) Calculate the volume of H_2S required to completely react with 20.0 cm³ of a solution of bismuth nitrate of concentration 0.0625 mol dm⁻³ at 25 °C and 1 atm pressure. [3]

(ii) What volume would be required if the temperature was doubled? [2]
(*1* mole of hydrogen sulfide occupies 24 dm³ at 25°C.)

Rhiannon's answer

(a) Mol HCl = 0.17 × 29.9 = 5.08 ✗ ①
Mol NaHCO₃ = 5.08 ✓②
Mol NaHCO₃ in original solution = 5.08 × 10 = 50.8 ✓
Mass in solution = 50.8 × 83.01 = 4217 g ✗ ③
% NaHCO₃ = $\dfrac{4.75}{4217}$ × 100 = 0.11% ✗

(b) (i) Mol Bi(NO₃)₃ = 0.0625 × 0.020 = 0.00125 ✓
Mol H₂S = 0.00125 × 2/3 = 0.000833 ✗ ④
Vol H₂S = 0.000833 × 24 = 0.0200 dm³ ✓②

(ii) $\dfrac{V_1}{T_1} = \dfrac{V_2}{T_2}$
$V_2 = \dfrac{V_1 T_2}{T_1} = \dfrac{0.020 \times 50}{25} = 0.040$ dm³ ✓✗ ⑤

Examiner commentary

① Rhiannon has forgotten to divide the volume by a thousand to change cm³ into dm³.
② Rhiannon gets the mark due to consequential marking.
③ The M_r of $NaHCO_3$ is incorrect.
④ Mole ratio should be 3/2.
⑤ Rhiannon has not converted °C into K.

Rhiannon achieves 5 out of 10 marks.

David's answer

(a) Mol HCl = 0.17 × 0.0299 = 5.08 × 10⁻³ ✓
Mol NaHCO₃ = 5.08 × 10⁻³ ✓
Mol NaHCO₃ in original solution = 5.08 × 10⁻³ × 10 = 5.08 × 10⁻² ✓
Mass in solution = 5.08 × 10⁻³ × 84.01 = 4.27 g ✓
% NaHCO₃ = $\dfrac{4.27}{4.75}$ × 100 = 89.9% ✓

(b) (i) Mol Bi(NO₃)₃ = 0.0625 × 0.020 = 1.25 × 10⁻³ ✓
Mol H₂S = 1.25 × 10⁻³ × 3/2 = 1.875 × 10⁻³ ✓
Vol H₂S = 1.875 × 10⁻³ × 24 = 0.045 dm³ ✓

(ii) $\dfrac{V_1}{T_1} = \dfrac{V_2}{T_2}$
$V_2 = \dfrac{V_1 T_2}{T_1} = \dfrac{0.045 \times 323}{298} = 0.0488$ dm³ ✓✓ ①

Examiner commentary

① David has done all the calculations correctly and he has used the correct units.

David achieves 10 out of 10 marks.

Equilibria

6 Sulfur dioxide and oxygen are heated at a certain temperature in a sealed container, until the following equilibrium is reached.

$$2SO_2(g) \ + \ O_2(g) \rightleftharpoons 2SO_3(g)$$

Analysis of this equilibrium mixture showed that it contained 0.042 mol SO_3, 0.038 mol SO_2 and 0.038 mol O_2. The total volume in the vessel was 2.0 dm³.

(a) State the meaning of the term *dynamic equilibrium*. [1]

(b) Write an expression for the equilibrium constant, K_c, giving its units. [2]

(c) Calculate the value of the equilibrium constant, K_c, at this temperature. [2]

(d) State and explain the effect of increasing the pressure on the equilibrium yield of SO_3. [2]

(e) When the temperature is increased the yield of SO_3 decreases. State, giving a reason for your answer, whether this reaction is endothermic or exothermic. [2]

Rhiannon's answer

(a) Dynamic equilibrium is when the forward reaction equals the backward reaction. ✗

(b) $K_c = \dfrac{(SO_3)^2}{(SO_2)^2(O_2)}$ mol⁻¹ dm³ ✗✓ ①

(c) $K_c = \dfrac{(0.042)^2}{(0.038)^2(0.038)}$ = 32.1 mol⁻¹ dm³ ✗✓ ②

(d) Increasing the pressure will shift the position of equilibrium to where there are least moles i.e. to the right hand side. ✓✗ ③

(e) The reaction is exothermic. To minimise the increase in temperature the equilibrium shifts to the left favouring the endothermic direction. ✓✓

Examiner commentary

① Square brackets must be used to show concentration.

② The total volume is 2 dm³, therefore the number of moles needs to be divided by 2 since concentration is mol dm⁻³.

③ Although Rhiannon's statement is correct, she has not answered the question fully – she has not stated the effect on the proportion of sulfur trioxide. So she only gains one mark.

Rhiannon achieves 5 out of 9 marks.

David's answer

(a) When the rate of the forward reaction equals the rate of the backward reaction. ✓

(b) $K_c = \dfrac{[SO_3]^2}{[SO_2]^2[O_2]}$ mol dm⁻³ ✓✗

(c) $K_c = \dfrac{(0.021)^2}{(0.019)^2(0.019)}$ = 64.3 mol dm⁻³ ✓✓ ①

(d) The yield would decrease because there are fewer (gas) moles on the right-hand side. ✓✓

(e) The reaction is exothermic since the equilibrium shifts to the left. ✓✗ ②

Examiner commentary

① Although the unit is incorrect, he does not lose a mark because he has already been penalised in part (b).

② David needs to explain why the equilibrium shifts to the left to gain both marks. Note that in a question of this type there is no mark for simply stating 'exothermic'. The marks are for the reasons.

David achieves 7 out of 9 marks.

Acid–base

7 (a) (i) Write an equation, including state symbols, for the reaction between magnesium carbonate and hydrochloric acid. *[1]*

(ii) Describe what you would see during this reaction. *[2]*

(b) Hydrochloric acid is referred to as a strong acid. Explain what is meant by the term *strong acid*. *[1]*

(c) Calculate the pH of an aqueous solution of hydrochloric acid of concentration 0.425 mol dm^{-3}. *[2]*

(d) Another strong acid, nitric acid can be reacted with ammonia to produce ammonium nitrate. The reaction can be represented by the equation:

$$NH_3(aq) + HNO_3(aq) \longrightarrow NH_4^+(aq) + NO_3^-(aq)$$

Explain why this is regarded as an acid–base reaction. *[2]*

Rhiannon's answer

(a) (i) $MgCO_3 + 2HCl \rightarrow MgCl_2 + H_2O + CO_2$ ✗ ①
    (ii) Bubbles of CO_2 forming. ✓✗ ②
(b) One that donates H$^+$ ions. ✗
(c) pH = $-\log[H^+]$ = $-\log 0.425$ = 0.37 ✓✓
(d) The nitric acid has donated a H$^+$ ion and the ammonia has accepted a H$^+$ ion. ✓✗ ③

Examiner commentary

① If the state symbols are asked for in a question, they must be given.

② There are 2 marks for this part, therefore two observations are required.

③ Although the statement is correct, Rhiannon has not answered the question fully. She has not explained the significance of the H$^+$ ion being donated or accepted.

Rhiannon achieves 4 out of 8 marks.

David's answer

(a) (i) $MgCO_3(s) + 2HCl(aq) \longrightarrow MgCl_2(aq) + H_2O(l) + CO_2(g)$ ✓
    (ii) A white solid reacts to form an aqueous solution and effervescence is seen. ✓✓①
(b) One that fully dissociates in an aqueous solution. ✓
(c) pH = 0.37 ✓✓②
(d) The nitric acid has donated a H$^+$ ion so is an acid. ✓
    The ammonia has accepted a H$^+$ ion so is a base. ✓

Examiner commentary

① Since this is a question about observation there is no need to name any substances to gain full marks.

② A correct answer gains full marks; however, if the answer was incorrect then David would receive 0 marks. It is always better to show your working as this can gain you marks if the final answer is incorrect.

David achieves 8 out of 8 marks.

Titration

8 Elinor determined the concentration of a solution of hydrochloric acid by titrating it against a standard solution of sodium carbonate. She rinsed the burette with acid, filled it to above the zero mark using a funnel, opened the tap and checked the burette jet. She removed the funnel and brought the acid level to exactly 0.00 cm^3.

She placed the sodium carbonate solution in a conical flask with an indicator and added the acid while swirling the flask. When the indicator gave signs of change, she added the acid drop by drop to the end-point.

The readings of her titrations were 20.80, 20.20, 20.05 and 20.10 cm^3 respectively.

(a) State why the burette was rinsed with the acid before filling. [1]

(b) State why the jet of the burette was looked at. [1]

(c) State why the funnel was removed. [1]

(d) State and explain whether there was any need for the acid level to be set exactly on zero. [1]

(e) State why the flask was swirled during the titration. [1]

(f) State why the acid was added drop by drop at the end. [1]

(g) Identify any anomalous result and calculate a mean value for her titration. [1]

Rhiannon's answer

(a) In case it was dirty. ✓

(b) To ensure that the acid flows freely. ✓

(c) For safety reasons. ✗

(d) Yes, to ensure that the acid level was the same at the beginning of each titration. ✗

(e) To mix all the reactants. ✓

(f) So that she would not add too much acid. ✓

(g) Mean value = $\dfrac{20.80 + 20.20 + 20.05 + 20.10}{4}$ = 20.29 cm^3 ✗ ①

Examiner commentary

① Rhiannon has included all four results in her calculation for the mean value. The first result is significantly higher than the others, therefore it should not be included in the calculation.

Rhiannon achieves 4 out of 7 marks.

David's answer

(a) To ensure that it was clean. ✓

(b) To check for air bubbles. ✓

(c) To prevent any further drops of acid falling in. ✓

(d) No, the result is the difference between the final and initial values. ✓

(e) To ensure that all the sodium carbonate reacts. ✓

(f) Not to overshoot the end-point. ✓

(g) Mean value 20.12 cm^3 ✓ ①

Examiner commentary

① Although David has not specifically stated that 20.80 cm^3 is an anomalous result, his answer for the mean titre shows that he has only included three titres in his calculation and so he gets the mark.

The question has been worded this way to indicate to the students that they should not use all the results in their calculation.

David achieves 7 out of 7 marks.

Bonding

9 (a) Use 'dot and cross' diagrams to show:

    (i)   Covalent bonding in Li_2,

    (ii) The formation of an ionic bond with LiCl.
       (Only outer electrons need to be shown.)   *[3]*

(b) Explain what is meant by a **co-ordinate bond** and give an example.   *[3]*

(c) List the forces of electrical attraction and repulsion that exist in a diatomic
covalent molecule.   *[3]*

Rhiannon's answer

(a) (i) Li ° × Li ✓ ①

   (ii) Li × °Cl° ⟶ Li × Cl ° ✓✗ ② ③

(b) A bond in which both electrons come from one of the atoms. ④

   H ° N × H ⁺ ×✓ ⑤

(c) Electron-electron repulsion, ✓ ⑥
Electrons attracted to both nuclei ✓ ⑦
⑧ ✗

Examiner commentary

① Correct drawing of Li_2.

② Electron transfer OK.

③ The formation of the ion with charges must be shown.

④ Description OK but limited.

⑤ Not clear that both electrons come from one atom.

⑥⑦ OK.

⑧ Nucleus – nucleus repulsion omitted.

Rhiannon gained 5 marks out of 9.

David's answer

(a) (i) Li ° × Li ✓ ①

   (ii) Li × °Cl° ⟶ Li⁺ × Cl°⁻ ✓✓

(b) A covalent-type bond where one of the atoms provided
both electrons in the electron bond pair. ✓ Such a bond
will be polar to some extent. ✓ ②

   H ° N ° H ⁺ ✓

(c) Electron-electron repulsion ✓
Electrons attracted to both nuclei ✓
Nucleus–nucleus repulsion ✓ ③

Examiner commentary

① Correct example used.

② Full answer and accurate diagram.

③ Correct – repulsion between the positive nuclei is often forgotten.

David obtained full marks.

Solid structures

10 (a) Explain what is meant by: (i) a giant molecule and (ii) a simple molecular structure. [2]

(b) Give an example of two types of giant molecule, excluding metals. [2]

(c) State the general physical properties of the two types of molecules in terms of (i) melting temperatures and (ii) solubility in water. [4]

Rhiannon's answer

(a) (i) A molecule containing a large number of atoms joined together,

(ii) A molecule containing a fairly small number of bonded atoms. ✓✓

(b) Diamond, polyethene ✓✓

(c) (i) Both have high m.t.s ✗✗ ①

(ii) Only simple molecules are soluble in water ✗✗ ②

Examiner commentary

Rhiannon clear on basic concepts but has little idea on the properties of these solids.

① Generally incorrect.

② Hydrocarbons are insoluble in water.

Rhiannon obtains 4 out of 8 marks.

David's answer

(a) (i) A molecule with a large number of bonded atoms.

(ii) A molecule having not many bonded atoms. ✓✓ ①

(b) Diamond, NaCl ✓✓ ②

(c) (i) Giants have high m.t.s while simple molecules are usually low ✓✓ ③

(ii) Many ionic giants are soluble while covalent giants are insoluble ✓✓

Examiner commentary

① Difficult in few words but OK.

② Good examples.

③ As in above but well put.

David understands the situation correctly and gives good examples. Shows a realisation that these statements are generalisations.

David obtains the full 8 marks.

Metals

11 (a) (i) State **two** properties that are characteristic of metals [2]

(ii) Relate these properties to the theory of metal structure. [3]

(b) Suggest why the melting temperatures of s-block elements fall down the Periodic Table; e.g., lithium 180 deg., sodium 98 deg. and caesium 29 deg. [1]

Rhiannon's answer

(a) (i) Metals are good electrical conductors. ✓✗ ①

(ii) Electrons are free to move around in metals so that their flow gives an electric current. ✓✗ ②

(b) The atoms are bigger. ✗ ③

Examiner commentary

① Only one property given.

② Again only electrical conductance dealt with.

③ A try but not specific enough.

Rhiannon weak here scoring only 2 marks but perhaps worth 3 altogether.

David's answer

(a) (i) Good electrical and thermal conductance, flexible and malleable. ✓✓ ①

(ii) Metal structures comprise a close-packed lattice of positive ions held together by a sea or gas of delocalised mobile electrons that gives good electrical conductance. The arrangement gives good flexibility since there are no fixed bond angles. ✓✓✓ ②

(b) This must be due to an increase in size down the group since they all have one electron involved in bonding and being further apart the ions are held less strongly. ✓ ③

Examiner commentary

① Sound and covering more properties than asked for.

② Very good answer showing mature understanding.

③ Suggestion is both sound and correct.

David gets full marks and shows his understanding.

Electronegativity

12 (a) Explain what is meant by the term electronegativity and show why electronegativity values are useful when considering bond polarity. *[4]*

(b) Using the electronegativity values below:

(i) Arrange the covalent bonds given in order of INCREASING polarity C–Cl; C–H; K–H; O–H. *[2]*

(ii) Write the δ+ and δ- charge signs of each atom in the bonds. *[2]*
Electronegativity values: K 0.8, H 2.1, C 2.5, Cl 3.0, O 3.5

Rhiannon's answer

(a) This is a measure of the attraction of an atom in a covalent bond to the electron pair in the bond. ✓ ✓ ①

Values are useful since the more electronegative elements will give more polar bonds ✗ ②

(b) (i) C±H‹C–Cl‹O–H‹H–K   ✓✗ ③

(ii) δ + δ-   δ+ δ-   δ+ δ-   δ- δ+

C--Cl   C---H   K--H   O--H   ✓✗ ④

Examiner commentary

① ② Definition OK, but it is the difference in electronegativity that is important not the actual value, e.g. F–F is nonpolar.

③ H–K should be before O–H

④ The C should be δ- in C–H

Rhiannon obtained 4 marks out of 8.

David's answer

(a) A measure of the ability of an atom in a covalent bond to attract the electron pair bond.✓✓ The values are useful since the difference between the two values for the atoms in the bond is proportional to the bond polarity. ✓✓ ①

(b) (i) C-H‹C-Cl‹H-K‹O-H   ✓✓ ②

(ii) δ+ δ-   δ- δ+   δ+ δ-   δ- δ+

C---Cl   C---H   K--H   O--H   ✓✓ ③

Examiner commentary

① All correct here, 4 marks awarded.

② Correct, so 2 marks given.

③ Correct so another 2 marks given.

David gained the full 8 marks.

Van der Waals bonding

13 (a) State which **one** of the following bonds is generally the weakest:
covalent   hydrogen   ionic   van der Waals *[1]*

(b) (i) Describe the nature of van der Waals forces and explain the
difference between the two types of van der Waals force. *[4]*

(ii) For each of these types above name a molecule where the force
is important. *[2]*

(c) State **one** effect that van der Waals forces have on the physical
properties of compounds. *[1]*

Rhiannon's answer

(a) Van der Waals ✓ ①

(b) (i) These are weak intermolecular forces
existing between all molecules. They are
electrical in nature and occur in all polar
molecules. ✓✓✗✗ ②

(ii) An example is that the forces are present in
liquid HI. ✓✗ ③

(c) The stronger the van der Waals force the higher
the boiling temperature of a liquid. ✓ ④

Examiner commentary

① Correct.

② Minimal answer but no mention of induced
dipole–induced dipole.

③ One example only.

④ OK.

Rhiannon is awarded 5 marks out of 8.

David's answer

(a) Van der Waals ✓

(b) (i) Weak intermolecular forces that exist between
all atoms and molecules. They are electrical in
nature and originate through the attraction
between opposite charges. In one type the
molecules are polar and have permanent
charge separation. In the second type, present
in all molecules, fluctuating dipoles induced by
electron movements come into phase to give an
attractive force between the molecules. ✓✓✓✓

(ii) An example of the first, dipole–dipole, type is
HI and of the second, induced dipole–induced
dipole type is He. ✓✓

(c) The boiling temperature of a liquid is governed by
the strength of the van der Waals force between the
molecules. ✓

Examiner commentary

Good answer with part (b) showing excellent
understanding.

David is awarded 8 marks out of 8.

Hydrogen bonding

14 Explain the nature of the hydrogen bond [4] and describe and explain its effect on the boiling temperatures of liquids containing hydrogen bonds [2] and on the solubility of compounds in water. *[2].*

Rhiannon's answer

The hydrogen bond is an intermolecular bond where hydrogen bonded to an electronegative element such as N, O or F bonds to a similar element in another molecule. ✓✓✓ ①

It is a very strong bond. ✗ ②

The boiling temperatures of liquids having hydrogen bonding are higher than expected ✓ ③ because the molecules in the liquid are held together more strongly so that more energy (i.e. a higher temperature) is needed to separate them. ✓ ④

Examiner commentary

①&② Correct, except that the H bond is only relatively strong compared with van der Waals bonding but is much weaker than covalent, ionic and metallic bonds.

③&④ Boiling temperatures sound.

✗ No mention of solubility.

Rhiannon achieves 5 out of 8 marks

David's answer

The hydrogen bond is an intermolecular bond where hydrogen bonded to an electronegative element such as N, O or F bonds to a similar element in another molecule. It is a stronger bond than a van der Waals bond but much weaker than a normal covalent bond. ✓✓✓✓

The boiling temperatures of liquids having hydrogen bonding are higher than expected because the molecules in the liquid are held together more strongly so that more energy (i.e. a higher temperature) is needed to separate them. ✓✓

Compounds such as lower alcohols having O-H bonds will be soluble in water through hydrogen bonding with the water. Many ionic solids are also soluble in water because of the interaction of the cation with the polar $O^{\delta-}$ and the anion with the $H^{\delta+}$ atoms in the water. ✓✓

Examiner commentary

A comprehensive answer. In the solubility part it is sensible to mention the solubility of ionic salts, although the interaction is perhaps not strictly hydrogen bonding in the normal sense.

Full marks are awarded.

Shapes of molecules

15 (a) Complete the table below by inserting the numbers of bonding pairs of electrons and naming the shapes of the molecules involved. *[4]*

Molecule	No. bonding pairs	No. lone pairs	Shape
$BeCl_2$		0	Linear
PCl_3	3	1	
CCl_4		0	

(b) State and explain the difference between bond pairs and lone pairs of electrons in governing the shapes of molecules. *[2]*

(c) The bond angles in CH_4, NH_3 and H_2O are 109.5, 107 and 104.5 degrees respectively. Explain this. *[3]*

Rhiannon's answer

(a) $BeCl_2$ 2 bond pairs ✓ ①
   PCl_3 trigonal planar shape ✗ ②
   CCl_4 4 bond pairs, tetrahedral shape ✓✓ ③

(b) Lone pairs repel more around the central atom ✓ ④ because they have opposite spin. ✗ ⑤

(c) Methane has no lone pairs, ammonia has one and water has two so that the bond angle is squeezed down along the series. ✓✓ ⑥

David's answer

(a) $BeCl_2$ 2 bond pairs
   PCl_3 trigonal pyramid shape
   CCl_4 4 bond pairs, tetrahedral shape ✓✓✓ ①

(b) Lone pairs have a larger repulsive effect around the central atom than bond pairs. This is because they are more localised around this atom while bond pairs are stretched out between the bonded atoms. ✓✓ ②

(c) Methane has four bond pairs only giving a symmetrical tetrahedral arrangement. In ammonia increased repulsion by the lone pair on the bond pairs closes up the H-N-H bond angle slightly and in water with two lone pairs the repulsion on the bond pairs is increased further giving a smaller H-O-H angle. ✓✓✓ ③

Examiner commentary

② Correct in ① and ③ but there is a lone pair in PCl_3 as in ammonia that distorts the planar shape into a trigonal pyramid.

④⑤ Lone pairs do repel more than bond pairs but this has nothing to do with spin but only that lone pairs are closer to the central atom.

⑥ Factually correct but limited and needs an explanation of why the angle is reduced.

Rhiannon has 6 marks out of 9.

Examiner commentary

① All are correct.

② Good explanation that brings out the essential difference.

③ Again a good explanation.

David has shown a good understanding for the full 9 marks.

Carbon structures

16 (a) Explain in terms of bonding and structure why diamond has a very high melting temperature. [2]

(b) Graphite also has a high melting temperature but is much softer than diamond. Explain this difference in terms of their structure. [2]

(c) Name a new form of carbon that is becoming technically important and briefly state what its structure is. [1]

Rhiannon's answer

(a) The carbon in diamond makes four strong covalent bonds ✓✗ ①

(b) Graphite has layers of hexagons with weak bonds between them. ✓✗ ②

(c) Graphene, like a layer of graphite. ✓ ③

Examiner commentary

① Correct but not enough for 2 marks.
② Basic idea OK but explanation weak.
③ Enough for what was asked.
Reasonable answers but limited in extent, 3/5.

David's answer

The four tetrahedral covalent bonds joining all the carbon atoms form a strong crosslinked structure. ✓✓ ①

Graphite forms giant-molecular hexagonal planes in two dimensions but these are weakly held together by van der Waals forces leading to softness and lubricating properties. ✓✓ ②

Graphene comprises single layers of graphite-type hexagons and has exceptional strength and other properties. ✓ ③

Examiner commentary

① Covers essential points.
② Relates structure to properties.
③ Gave more than was asked for; good to impress examiner.
Good throughout – full marks.

Periodic Table

17 (a) (i) State the meaning of the term electronegativity, describe how its value changes across and down the Periodic Table and explain this trend. *[3]*

(ii) Explain why both electronegativity and ionisation energy values change in the same way in going across and also in going down the Periodic Table. *[1]*

(iii) Explain why the melting temperatures of the halogen elements increase down the group. *[2]*

Rhiannon's answer

(a) (i) A measure of the electron-attracting power of an atom in a covalent bond. ✓ Its value increases across the Periodic Table and down a group. ✓✗ ①

(ii) Because they both increase diagonally up the Periodic table. ✗ ②

(iii) The van der Waals forces increase down the group as the number of electrons increases. ✓✗ ③

Examiner commentary
① Correct definition but the electronegativity decreases down a group.
② Not an explanation.
③ Partly correct but incomplete as an explanation.
Rhiannon achieves 3 marks out of 6.

David's answer

(a) (i) A measure of the electron-attracting power of an atom in a covalent bond. Its value increases across the Periodic Table but decreases down a group, so effectively increasing diagonally upwards across the Table. ✓✓✓ ①

(ii) Electronegativity reflects the pull of an atom on its electron in a bond. This is essentially what ionisation energy is so that both will follow the same trends. In fact one way of determining electronegativity includes ionisation energy as a major contributor. ✓ ②

(iii) The halogens comprise covalent diatomic molecules held together in the solid by van der Waals forces. The van der Waals forces increase down the group as the number of electrons increases so that the m.t.s increase. ✓✓ ③

Examiner commentary
① All correct and clearly put.
② Gives an extra point beyond the Specification that is in fact vey helpful.
③ Full and correct answer.
David gives a very good answer.

Periodic Table trends

18 (a) Describe and explain the general change in ionisation energies:

    (i) Across a period, e.g. from Na to Ar.

    (ii) Down a group, e.g. From Li to Cs. [4]

  (b) (i) Give one example of a (I) basic oxide, (II) acidic oxide. [2]

    (ii) State the regions of the Periodic Table in which the elements form (I) basic and (II) acidic oxides. [2]

Rhiannon's answer

(a) (i) IEs increase across a period because there is a steady increase in the number of protons in the nucleus. ✓✓ ①

  (ii) IEs fall down a group because the outer electron is further from the nucleus in the larger atoms. ✓✗ ②

(b) (i) (I) MgO, (II) SO₂ ✓✓ ③

  (ii) (I) The LHS, (II) the RHS. ✓✗ ④

David's answer

(a) (i) IEs increase across a period because there is a steady increase in the number of protons in the nucleus while the ionisable electron is in the same orbital and electron shielding is not much increased. ✓✓ ①

  (ii) IEs decrease slightly down a group since the increase in nuclear charge is outweighed by increased screening by the filled orbitals so that the effective nuclear charge decreases. ✓✓

(b) (i) (I) CaO, (II) NO₂ ✓✓

  (ii) (I) The LHS with the bottom of the s-block being most basic. ✓

    (II) The RHS, especially the upper parts of groups 5 and 6. ✓②

Examiner commentary

① Should have added 'without much increase in electron shielding'.

② This explanation itself is rather unsound (see David's answer for a better explanation of this), and the distance from the nucleus is a secondary factor controlled by the effective nuclear charge.

③ This is basically acceptable.

④ Incorrect. She should have referred to regions of the LHS and RHS.

Rhiannon scored 6 out of the 8 possible marks.

Examiner commentary

① Good answers showing clear understanding of the factors.

② Scores all the marks but the actual state of group 7 oxides is not in the spec. and inert gases do not form oxides.

David scored the full 8 marks.

Oxidation numbers

19 (a) State four of the rules that are used to assign oxidation numbers to elements in compounds. *[4]*

(b) Evaluate the oxidation numbers of all of the atoms in the following compounds or ions:
$CaSO_4$, F_2, Na_2CO_3, NH_4^+ *[4]*

Rhiannon's answer

(a) Elements are zero, oxygen is –2, hydrogen is 1, in ions the oxidation number is the charge on the ion. ✓✓✓✗ ①

(b) Ca is 2, O is –2, S is 6; F is 0, Na is 1, O is –2, C is 4; H is 1, N is 4 ✓✓✓✗ ②

Examiner commentary

① Three correct – can use 2 or +2 or II for positive numbers and -3 or -III for negative. The oxidation number is only the charge on the ion for atoms, such as +1 for Na+ and not for ionic compounds. Do not write 2+ or 3- for oxidation numbers, these are for ionic charges only.

② All correct except for NH_4^+ where sum of N and all the H oxidation numbers must be +1 therefore N is -3 and H4 is 4 times +1.

Rhiannon gets 6 out of a possible 8 marks.

David's answer

(a) Uncombined elements are 0, in simple ions the oxidation number is the charge on the ion, hydrogen is +1, the sum of the oxidation numbers in an ion equals the charge on the ion or in an uncharged compound equals 0. ✓✓✓✓ ①

(b) Ca is 2, O is -2, S is 6; F is 0, Na is 1, O is -2, C is 4; H is 1, N is -3 ✓✓✓✓ ②

Examiner commentary

① All correct including the good point about the sums of the oxidation numbers.

② Correct, with the compound ion dealt with successfully.

David scores 8 out 8.

s-Block elements

20

(a) Write a balanced equation for the reaction of calcium metal with hydrochloric acid. [1]

(b) State how group I and group II elements compare with regard to:

(i) Their reactivity with water and oxygen

(ii) The solubility of their salts. [4]

(c) State also how the solubilities of group II hydroxides and sulfates change in descending the group. [2]

Rhiannon's answer

(a) $Ca + HCl = CaCl + H_2$ ✗ ①

(b) (i) Both groups react with water to give hydroxides or oxides and with oxygen to give oxides. ✓ ②

(ii) All group I salts are soluble in water ✓ and all group II salts are insoluble. ✗ ③

(c) Hydroxides become more soluble down group II; sulfates become less soluble down group II. ✓✓ ④

Examiner commentary

① Incorrect; errors in balancing and valency are very common and basics must be memorised.

② OK but limited answer.

③ Some group II salts are soluble, such as nitrates and halides.

④ This is generally acceptable.

Rhiannon scored 4 out of 7.

David's answer

(a) $Ca + 2HCl = CaCl_2 + H_2$ ✓ ①

(b) (i) Both groups react with water to give hydroxides or oxides and with oxygen to give oxides. Group I elements are more reactive than those of group II and in both cases reactivity increases down the group. ✓✓

(ii) All group I salts are soluble in water; in group II nitrates and halides are usually soluble, sulfates and hydroxides may or may not be soluble and carbonates are not very soluble. ✓✓ ②

(c) Hydroxides become more soluble down the group whereas sulfates become less soluble. ✓✓ ③

Examiner commentary

① Correct.

② Good clear answers in both parts.

③ Correct.

David has full marks.

The halogens

21 (a) (i) State what is meant by a **displacement reaction** in the halogens.

(ii) Explain why such reactions occur.

(iii) Write a balanced equation for the reaction of chlorine with potassium iodide solution. *[3]*

(b) (i) State how you would determine whether a solution of a halide in dilute nitric acid contains chloride, bromide or iodide ions and what would be observed. *[2]*

(ii) Explain why this test is useful in organic as well as inorganic chemistry and describe the additional steps needed in the organic case. *[2]*

Rhiannon's answer

(a) (i) Reactions in which one halogen exchanges places with another halogen in a halide. ✓ ①

(ii) Halogens higher in the Periodic Table are stronger oxidising agents. ✓

(iii) $Cl_2 + 2KI = I_2 + 2KCl$ ✓ ②

(b) (i) Add silver nitrate solution when a precipitate would form. ✓✗ ③

(ii) To identify the halogen in the organic compound: ✓ the organic halide must be heated. ✗ ④

Examiner commentary

① Correct but (i) and (ii) rather terse.

② OK.

③ Nothing about distinguishing the halogens.

④ Nothing useful in the second part here.

Rhiannon scored 5 out of 7.

David's answer

(a) (i) Reactions in which one halogen exchanges places with a lower halogen in a halide.

(ii) Halogens higher in the Periodic Table are stronger oxidising agents and thus remove an electron from the halide to liberate the free halogen while being reduced to the halide.

(iii) $Cl_2 + 2KI = I_2 + 2KCl$ ✓✓✓

(b) (i) Add silver nitrate solution when a precipitate of silver halide forms. Silver chloride only dissolves in dilute ammonia, identifying chloride and silver iodide is yellow in colour as against the buff colour of bromide. ✓✓

(ii) This test is useful in identifying the halogen present in halogenoalkanes but the halogen must first be liberated from the organic compound by warming with NaOH solution to hydrolyse it and liberate the halide ion. The solution must then be acidified with nitric acid so that only silver halide is precipitated when silver nitrate is added. ✓✓

Examiner commentary

Very clear and confident answers here with some additional material to impress the examiner.

David obtained full marks.

Enthalpy (theory)

22 (a) State Hess's law. [1]

(b) Define the term molar standard enthalpy change of combustion. [2]

(c) Butan-1-ol burns in air according to the equation:

$$C_4H_9OH \text{ (l)} + 6O_2\text{(g)} \longrightarrow 4CO_2\text{(g)} + 5H_2O\text{(l)}$$

(i) I Calculate the enthalpy change for this reaction using the following enthalpy changes of formation, $\Delta_f H^\theta$. [2]

Compound	C_4H_9OH (l)	O_2(g)	CO_2(g)	H_2O(l)
$\Delta_f H^\theta$ / kJ mol^{-1}	−327	0	−394	−286

II State why O_2(g) has a value of zero for its standard enthalpy change of formation. [1]

(ii) Calculate the enthalpy change for this reaction using the following average bond enthalpy values, E. [3]

Bond	C − C	C − H	C − O	C = O	O − H	O = O
E / kJ mol^{-1}	348	413	360	805	463	496

(iii) Which calculation is more accurate? Justify your answer. [1]

Rhiannon's answer

(a) The enthalpy change is independent of the route. ✓

(b) The enthalpy change when I mole of a substance is completely burned in oxygen. ✓✗ ①

(c) (i) I $\Delta H = -2679$ kJ mol^{-1}. ✓✓

II Because it's an element. ✗ ①

(ii) Bonds broken 4(C − C)+9(C − H)+
(C − O)+(O − H)+5(O = O) =
8908 kJ mol^{-1} ✗ ②

Bonds formed 8(C = O)+12(O − H) =
−11996 kJ mol^{-1} ✓

$\Delta H = 8908 - 11996 = -3088$ kJ mol^{-1}. ✓ ③

(iii) Part (i) because part (ii) uses averages not the exact value. ✓

Examiner commentary

① To get full marks Rhiannon must add 'under standard conditions'.

② Always draw out each molecule so that you can see the bonds broken and the bonds made.

③ Rhiannon gets the mark for 'error carried forward'.

Rhiannon achieves 7 out of 10 marks.

David's answer

(a) The enthalpy change of a reaction is independent of the route taken from the reactants to the products. ✓

(b) The enthalpy change when a substance is completely burned in oxygen under standard conditions. ✓✗①

(c) (i) I $\Delta H = 4(-394) + 5(-286) - (-327)$
$= -2679$ kJ mol^{-1}. ✓✓

II It's an element under standard conditions. ✓

(ii)

```
    H H H H                          O
    | | | |                         ⁄ \
H - C-C-C-C-O-H + 6O=O  →  4O=C=O + 6H   H
    | | | |
    H H H H
```

Bonds broken 3(C − C) = 1044   9(C − H) = 3117
1(C − O) = 360   1(O − H) = 463
6(O = O) = 2976   Total = 8560 ✓

Bonds formed 8 (C = O) = −6440   12(O − H) = −5556
Total = −11996 ✓②

$\Delta H = 8560 - 11996 = -3436$ kJ mol^{-1}. ✓

(iii) The calculation using enthalpy change of formation since average bond enthalpy used not actual ones. ✓

Examiner commentary

① David needs to state '1 mole of a substance'.

② Remember that bond breaking is always endothermic (+) and bond forming exothermic (−).

David achieves 9 out of 10 marks.

Enthalpy (practical)

23 Describe a laboratory experiment of your choice, for determining the enthalpy change of a reaction. Your answer should include details of the apparatus to be used, the measurements to be taken and the way in which you would use your results to determine the enthalpy change. *[6]*

Rhiannon's answer

To determine the enthalpy change for the neutralisation reaction between magnesium oxide and dilute hydrochloric acid.

Pour hydrochloric acid into a coffee cup, take the temperature then add a weighed amount of magnesium oxide. Take the temperature of the solution every 30 seconds for the next 6 minutes. ✓✓✗✗ ①

To find the enthalpy of the reaction use the expressions $q = mc\Delta T$ and $\Delta H = -q/n$

where m is the mass of the solution, c is a given constant, ΔT the maximum temperature change and n the number of moles of magnesium oxide. ✓✗ ②

David's answer

I will determine the enthalpy change for the displacement reaction between zinc and copper(II) sulfate solution using the following method.

- Pipette 50.0 cm³ of copper(II) sulfate solution of known concentration into a polystyrene cup.
- Put a thermometer through the hole in the lid and record the temperature to the nearest 0.1 °C every half minute until the reading is constant. ✓✓
- Weigh about 6 g of zinc and add the powder to the cup. ①
- Stir the solution well and record the temperature every half minute until the temperature fall is constant. ✓✓
- Draw a graph of temperature against time and extrapolate the curve to the beginning of experiment to find the maximum temperature change.
- Use $q = mc\Delta T$ to find the enthalpy change for the experiment
- Scale to 1 mole by dividing by the number of moles of copper(II) sulfate (the substance that is not in excess). ✓✓ ②

Examiner commentary

① Although the method is correct, Rhiannon's answer is not detailed enough. She needs to:

- State what volume of hydrochloric acid she's adding and what apparatus she uses to add it.
- Ensure that the temperature of the acid is constant before she adds the magnesium oxide and state the accuracy of the thermometer.
- Stir the reactants vigorously and ensure that the coffee cup has a lid.

② To obtain ΔT she needs to draw a graph of temperature against time and extrapolate to the beginning of the experiment. She also needs to make it clear that if she is using the number of moles of magnesium oxide then the acid is in excess.

Rhiannon achieves 3 out of 6 marks.

Examiner commentary

① David did not need to weigh the zinc accurately because it was in excess and the m in the expression $q = mc\Delta T$ refers to the mass of the solution.

② David has an excellent grasp of this topic and has given a full description of the experiment. He could have stated what the m and c stand for in the expression for q. However, in a 6-mark question there will be more marking points than marks available in an indicative content and the marks will be banded into three groups. As long as all key elements of the indicative content have been included, he can gain full marks.

David achieves 6 out of 6 marks.

Reaction rates (theory)

24 Discuss and explain the following by considering the movement and the energy of the particles involved.

(a) Calcium carbonate reacts faster with:

    (i)   concentrated hydrochloric acid than it does with dilute hydrochloric acid   [3]

    (ii)   dilute hydrochloric acid at 60 °C than at 20 °C.   [4]

(b) Gaseous ethene and hydrogen react faster with each other when nickel powder is present.   [3]

Rhiannon's answer

(a) (i) When the hydrochloric acid is concentrated, there are more molecules present in a certain volume. ✓ This means there is more chance of collisions per unit time and so the reaction goes faster. ✓✗ ①

   (ii) The different temperatures mean that the particles in the acid will be moving at different speeds, the higher the temperature the higher the speed and their energy increase. ✓ If the particles have greater energy this means there will be more successful collisions per second. ✓ A greater rate of collisions means a faster rate of reaction. ✗ ②

(b) Nickel acts as a catalyst, so speeds up the reaction. ✓ A catalyst lowers the activation energy ✗ ③ so more particles have enough energy to react and the rate of reaction is faster. ✓

Examiner commentary

① Rhiannon's answer gives the impression that all collisions cause a reaction – only molecules with sufficient energy react when they collide. This is a common mistake – be careful to avoid it.

② Rhiannon's idea that the reaction goes faster because there are more collisions is not detailed enough to gain any more marks. She should have used the idea of activation energy – the minimum energy needed by a particle in order to react on collision – in her answer.

③ A catalyst does not lower activation energy: it provides a different route of lower activation energy.

Rhiannon achieves 6 out of 10 marks.

David's answer

(a) (i) In concentrated hydrochloric acid there are more particles in a given volume ✓ therefore there will be more molecular collisions taking place, so there will be a greater frequency of collisions ✓ ① that actually cause a reaction. ✓ This makes the reaction go faster.

   (ii) In order to react, collisions must occur between particles with at least a minimum amount of energy – the activation energy. ✓ When the temperature of the acid increases, the kinetic energy of its particles also increases. ✓ This results in a larger proportion of the particles having enough energy to react ✓ so the frequency of successful collisions increase and the rate of reaction increases. ✓ ②

(b) Nickel acts as a catalyst, and isn't used up in the reaction. ✓ The nickel provides an alternative route for the reaction, with a lower activation energy. ✓ This results in a greater proportion of particles having enough energy to react and the reaction proceeds more quickly. ✓ ③

Examiner commentary

① Frequency of collision is equivalent to collisions per unit time.

② David has given a full explanation expected at this level.

③ This is a full and suitable answer. David could have drawn a labelled enthalpy profile diagram or distribution curve to obtain 2 marks, but an explanation would be needed for the third mark.

David achieves 10 out of 10 marks.

Reaction rates (practical)

25 Hydrogen peroxide, H_2O_2, decomposes to give water and oxygen.
An experiment was conducted to investigate the decomposition of 40 cm³ of
0.40 mol dm⁻³ H_2O_2 solution at a constant temperature of 25 °C in the presence
of 0.5 g of MnO_2 catalyst. The results are shown below.

Time /s	0	10	20	40	80	120	160	200	240
Amount O_2 / mol	0	0.00026	0.00050	0.00094	0.00154	0.00184	0.00196	0.00200	0.00200

(a) Outline a suitable method, including essential apparatus, for carrying
out an experiment to obtain these results. State how you would
calculate the number of moles of oxygen from your experimental results. *[6]*

(b) Use the results to plot a graph of the amount of oxygen against time
and from the graph calculate the initial rate of reaction. *[5]*

(c) Sketch on your graph the curve you would expect to obtain if the
reaction were repeated using 40 cm³ of 0.20 mol dm⁻³ H_2O_2 solution.
Explain any differences in the curves. *[3]*

Rhiannon's answer

(a) Using a measuring cylinder pour the H_2O_2 solution into a conical flask ✓ add the
MnO_2 and measure the volume of oxygen formed every 10 seconds ✓ until the
reaction stops using a stopwatch. ✗ ①
To change volume into moles, I would divide each volume by 22 400 because I mole
of gas occupies 22.4 dm³ at standard temperature and pressure. ✓✗ ②

(b) 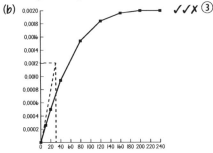 ✓✓✗ ③

Initial rate = $\dfrac{moles}{time}$ = $\dfrac{0.0012}{30}$ = 0.00004 ✓✗ ④

(c) 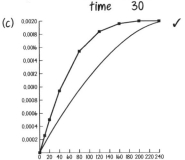 ✓

The curve is less steep so this shows that the reaction is slower. ✗✗ ⑤

Examiner commentary

① The rest of the method is
not detailed enough to score
any marks. Using a labelled
diagram to show the set-up
of the apparatus can gain you
marks.

② Rhiannon scores 1 mark
for recognising that the moles
of oxygen could be obtained
from the volume. She fails
to score the second mark
because the value 22.4 dm³ is
the volume occupied at 0 °C
not 25 °C.

③ Rhiannon has not labelled
the axes and so loses 1 mark.

④ Rhiannon has not given a
unit for the rate and so loses
1 mark.

⑤ Rhiannon scores 1 mark
for the curve being less steep
but she has not explained
why it is less steep.

**Rhiannon achieves 7 out of
14 marks.**

David's answer

(a) A measuring cylinder would be used to pour 40 cm³ of the H_2O_2 solution into a conical flask. ✓ 0.5 g of the MnO_2 would be weighed accurately. The reaction is carried out at a constant temperature of 25 °C. ✓ When ready, the MnO_2 would be added quickly to the H_2O_2 solution and a bung and delivery tube attached to a gas syringe inserted into the flask ✓ and a stopwatch started. At appropriate intervals the volume of the oxygen in the syringe would be recorded. ✓①

To calculate the number of moles of oxygen I would use the ideal gas equation $PV = nRT$. ✓   This can be rearranged to $n = \dfrac{PV}{RT}$ where P is in Nm^{-2}, V in m^3,

T in K and R is the gas constant. ✓②

(b)

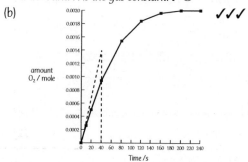

✓✓✓

Initial rate = $\dfrac{moles}{time}$ = $\dfrac{0.0014}{40}$ = 3.5×10^{-5} mol s⁻¹ ✓✓③

(c)

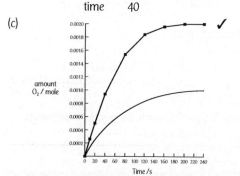

✓

The curve is less steep because there is a decrease in concentration so there are fewer successful collisions per unit time. ✓ The amount of oxygen has halved since the number of moles of hydrogen peroxide has halved. ✓

Examiner commentary:

① David's choice of apparatus and method is good and this gains him 4 marks. He could have improved his method by stating how he would keep the temperature at 25 °C.

② It is important that the correct units are used in this expression. You are not expected to know the value for atmospheric pressure or the gas constant – these will be given on an exam paper.

③ Although the gradient is different from Rhiannon's answer, examiners are aware that small variations will happen when graphs are drawn so a mark is awarded for an answer within a sensible range.

David achieves 14 out of 14 marks

The wider impact of chemistry

26 In addition to carbon dioxide the combustion of some fuels such as diesel and kerosene in vehicles and planes produces harmful oxides of nitrogen (NO_x) and is raising questions over the advantage of diesel in cars.

(a) (i) Write an equation to show the moles of CO_2 and water produced from one mole of diesel (C_8H_{18}). *[1]*

(ii) Diesel cars average 60 miles per gallon of fuel as against 40 miles with petrol engines. Atmospheric CO_2 levels are now increasing by 0.5% per year. If all engines ran on diesel as against petrol and this was the only factor in the increase, estimate what this value would then be. *[1]*

(b) (i) If NO_x is written as NO_2 write a balanced equation for the reaction of air in the engine to form NO_2. *[1]*

(ii) The emission of NO_x can be reduced by adding NH_3 to the exhaust to reform nitrogen. Write a balanced equation for this reaction. *[1]*

(iii) Give the oxidation states for all the atoms in this equation. *[2]*

(c) Suggest a reason why the gas NO_x is harmful. *[1]*

Rhiannon's answer

(a) (i) $C_8H_{18} + O_2 = 8CO_2 + 9H_2O$ ✓ ①
   (ii) no answer ✗ ②

(b) (i) $N_2 + 2O_2 = 2NO_2$ ✓
   (ii) $NH_3 + NO_2 = N_2 + 2H_2O$ ✗ ③
   (iii) LHS N is +3 and +4, O is −2; RHS N is 0, H is +1 and O is −2. ✗

(c) It is bad to breathe. ✓ ④

Examiner commentary

① Equation OK, no need to balance.
② Did not understand the question.
③ Equations not balanced.
④ Limited but essentially correct.

Rhiannon has trouble in balancing equations and oxidation numbers and understanding question (a) (ii). She scored three marks.

David's answer

(a) (i) $C_8H_{18} + O_2 = 8CO_2 + 9H_2O$ ✓
   (ii) 0.75% pa ✗ ①

(b) (i) $N_2 + 2O_2 = 2NO_2$ ✓
   (ii) $4NH_3 + 3NO_2 = 7/2N_2 + 6H_2O$ ✓
   (iii) LHS N is −3 and +4, H is 1, O is −2; RHS N is 0, H is 1 and O is −2 ✓✓ ②

(c) This acidic corrosive gas is dangerous for the lungs and breathing passages. ✓ ③

Examiner commentary

① (a)(ii) Got calculation upside down but understood principle.
② Sound on all aspects of balancing equations and oxidation states.
③ Full explanation well worth the mark.

Good answers here with slight slip in calculation that is easy to make. Six marks.

Hydrocarbons and *E-Z* isomerism

27 (a) Gas oil is a hydrocarbon fraction obtained from petroleum.

　　(i) State how gas oil and other hydrocarbon fractions are obtained, starting from petroleum. *[1]*

　　(ii) State why some of the gas oil fraction is cracked. *[1]*

(b) Tridecane, $C_{13}H_{28}$, is one of the compounds present in gas oil.

One of the equations used to represent the cracking of tridecane is shown below.

$$C_{13}H_{28} \rightarrow \text{Compound Z} + C_4H_6 + H_2$$

　　(i) Find the molecular formula of compound Z by using the equation. *[1]*

　　(ii) Write the molecular formula of a compound which is in the same homologous series as compound Z but contains **six** carbon atoms per molecule. *[1]*

(c) Another of the products made by cracking tridecane is but-1,3-diene.

$$\underset{\quad}{H_2C}=\overset{\overset{\displaystyle H}{|}}{C}-\overset{\overset{\displaystyle H}{|}}{C}=CH_2$$

But-1,3-diene reacts with bromine to form several products.

　　(i) One of the products is 3,4-dibromobut-1-ene, $CH_2{=}CH{-}CHBr{-}CH_2Br$.

A possible mechanism for this bromination is shown below.

$$H_2C=\overset{\overset{\displaystyle H}{|}}{C}-\overset{\overset{\displaystyle H}{|}}{\underset{\underset{\displaystyle \underset{\delta^-Br}{|}}{\delta^+Br}}{C}}{\Large\nparallel} CH_2 \longrightarrow H_2C=\overset{\overset{\displaystyle H}{|}}{C}-\overset{\overset{\displaystyle H}{|}}{\underset{\underset{\displaystyle +}{}}{C}}-\overset{\overset{\displaystyle Br}{|}}{C}H_2 \qquad \text{carbocation A}$$

$$:Br^-$$

$$\downarrow$$

$$H_2C=\overset{\overset{\displaystyle H}{|}}{C}-\overset{\overset{\displaystyle H}{|}}{\underset{\underset{\displaystyle Br}{|}}{C}}-\overset{\overset{\displaystyle Br}{|}}{C}H_2$$

　　I State what is represented by the curly arrow. *[1]*

　　II State what is represented by the δ+ and δ- symbols on the bromine atoms. *[1]*

　　III The mechanism shows the formation of a carbocation **A**.

Explain why the mechanism is less likely to proceed via carbocation **B**. *[1]*

$$H_2C=\overset{\overset{\displaystyle H}{|}}{C}-\overset{\overset{\displaystyle H}{|}}{\underset{\underset{\displaystyle H}{|}}{C}}-\overset{\overset{\displaystyle Br}{|}}{\underset{\underset{\displaystyle H}{|}}{C}}^+ \qquad \text{carbocation B}$$

> (ii) Another product of the bromination of but-1,3-diene is
> 1,4-dibromobut-2-ene,
> $BrCH_2–CH=CH–CH_2Br$.
> This shows E-Z isomerism.
>
> I Draw the structure of 1,4-dibromobut-2-ene to show the Z isomer. *[1]*
>
> II Explain why 1,4-dibromobut-2-ene shows E-Z isomerism. *[1]*

Rhiannon's answer

(a) (i) Distillation. ✗ ①

(ii) Cracked to make smaller, more useful hydrocarbons. ✓ ②

(b) (i) C_9H_{18} ✗ ③

(ii) C_6H_{12} ✓

(c) (i) I Curly arrows show electron movement. ✗ ④

II The $\delta+$ and $\delta-$ show the presence of a dipole. ✓

III Carbocation A is more stable. ✗ ⑤

(ii) BrCH₂ C=C CH₂Br ✓ ⑥ (with H, H)

II The compound has a double bond. ✗ ⑦

Examiner commentary

① 'Distillation' alone is never acceptable when the answer' fractional distillation' is required.

② An acceptable answer. Rhiannon does state that the molecules are smaller and why this is important.

③ Rhiannon actually makes a mistake here in the formula of compound Z. However, once she has given the formula of an alkene in (i) then it follows that the formula in (ii) should also be an alkene. This is an 'error carried forward' and so she scores the mark for (ii).

④ This is a good example of why precise terminology is needed. It is the actual movement of a pair of electrons that are shown by the curly arrow. Rhiannon's answer is not wrong but is imprecise.

⑤ As in I, it is the detail that means Rhiannon's answer does not score any marks. What is there about the carbocation A that makes it more stable? She could alternatively have said that it is a secondary carbocation.

⑥ Rhiannon is correct. It's a good idea if you can think of some hint or tip that shows you a way of remembering which is which!

⑦ Once again Rhiannon's answer is not wrong but it does not score since it does not explain the significance of the double bond in the context of the existence of the two isomers.

Rhiannon scores 4 out of 9 marks.

David's answer

(a) (i) Fractional distillation – separates according to the different boiling points of the fractions. ✓ ①

(ii) It creates smaller molecules that can be used, for example, in petrol. ✓ ②

(b) (i) C_9H_{20} ✓

(ii) C_6H_{14} ✓

(c) (i) I The curly arrow shows the movement of a pair of electrons. ✓ ③

II $\delta+$ shows that end of the molecule is slightly positive and $\delta-$ shows that end of the molecule is slightly negative. ✓

III Carbocation A is more stable because it has 2 C atoms attached to the C atom with the +. ✓

(ii) I H C=C CH₂Br ✗ ④ (with BrCH₂, H)

II There is no rotation about a carbon to carbon double bond. ✓

Examiner commentary

① From the wording of question, the answer 'fractional distillation' alone might have been sufficient but David makes sure of the mark by explaining the meaning of the term.

② Although David takes a different approach from Rhiannon here, both answers are acceptable – both highlight the fact that the molecules are smaller and both explain, in a different way, the significance of this.

③ David accurately describes the type of movement of the pair of electrons that are shown by the curly arrow.

④ Incorrect –David has drawn the E form.

David scores 8 out of 9 marks.

Calculation of formulae and polymers

Q&A 28

(a) Compound **A** contains carbon, hydrogen and oxygen only. It has a molar mass of 88.2g mol^{-1}. Quantitative analysis of the compound shows that its percentage composition by mass contains 54.5% carbon and 9.10% hydrogen.

Calculate both the empirical and molecular formulae of compound **A**. *[4]*

(b) (i) Propan-1-ol can be completely oxidised to form compound **B**.

Name compound **B** and write the equation to show this oxidation. You may use [O] to represent the oxidising agent. *[3]*

(ii) Propan-1-ol can also form propene by a dehydration reaction. Name a suitable reagent for this reaction. *[1]*

(c) Propene can be polymerised to form poly(propene). Give the formula of the repeating unit in poly(propene). *[1]*

(d) Substituted alkenes can also be polymerised to give useful polymers. Name an important polymer formed from a substituted alkene. *[1]*

Rhiannon's answer

(a) % oxygen = 36.4 ✓

$$C : H : O$$
$$= \frac{54.5}{12} : \frac{9.10}{1.01} : \frac{36.4}{16}$$
$$= 4.54 : 9.01 : 2.28 \qquad ✓$$
$$= 5 : 9 : 2$$

Empirical formula = $C_5H_9O_2$ ✗

Empirical M_r = 101

Molecular formula = ✗ ①

(b) (i) Propanoic acid ✓

$C_3H_7OH + [O] \rightarrow C_3H_7COOH$ ✗✗ ②

(ii) sulfuric acid ✓ ③

(c)

$$\begin{array}{c} CH_3 \quad H \\ | \qquad | \\ -C-C-H \\ | \qquad | \\ H \quad H \end{array} \qquad ✗ ④$$

(d) PVC ✓ ⑤

Examiner commentary

① Rhiannon uses the percentages to calculate the number of moles of each element, but she then approximates these to whole numbers. In exam questions the numbers will always produce obvious whole numbers at this stage and you should never approximate. Rhiannon might have realised she had made a mistake, and gone back and corrected it, when her empirical formula M_r was not related directly to the value given in the question.

② Rhiannon named the acid correctly but did not recognise that one of the carbons in the C_3H_7 group is needed to form the COOH in the acid.

③ A variety of acceptable dehydrating agents exist. Rhiannon's answer, sulfuric acid, was accepted although concentrated sulfuric acid would have been better.

④ In drawing the formula for the repeat unit of a polymer the end bonds must show that the chain continues, i.e. have nothing attached to them.

⑤ There are generally a wide variety of answers to questions that ask for names or uses of particular groups of compounds. It is important to note **important** and not quote lab type/ small scale uses.

Rhiannon scores 5 out of 10.

David's answer

(a) % oxygen = 36.4 ✓

$$\quad\quad C \ : \ H \ : \ O$$

$$= \ \frac{54.5}{12} \ : \ \frac{9.10}{1.01} \ : \ \frac{36.4}{16}$$

$$= \ 4.54 \ : \ 9.01 \ : \ 2.28 \quad ✓$$

$$= \ 1.99 \ : \ 3.95 \ : \ 1$$

Empirical formula = C_2H_4O ✓

Molecular formula = $C_4H_8O_2$ ✓ ①

(b) (i) Propanoic acid ✓

$\quad\quad C_3H_7OH + [O] \rightarrow C_2H_5COOH + H_2O$ ✓✗ ②

(ii) Aluminium oxide ✓

(c)
$$\left[\begin{array}{c} CH_3 \ \ H \\ | \quad\; | \\ C - C \\ | \quad\; | \\ H \quad H \end{array} \right]$$ ✓ ③

(d) PTFE ✓ ④

Examiner commentary

① David correctly uses the percentages to calculate the number of moles of each element, and then divides by the smallest to obtain a whole number ratio. Although he scored full marks, he might have been more secure in this if he had shown his calculation of the M_r of his empirical formula.

② David named the acid correctly and knew that one of the carbons in the C_3H_7 group is needed to form the COOH in the acid. While he scored the mark for this and realised that water was the other product, he did not actually balance the equation.

③ The end bond shows that the chain continues, i.e. has nothing attached.

④ One of a large variety of acceptable answers.

David scores 9 out of 10.

Halogenoalkanes

29 (a) Methane reacts with gaseous chlorine giving chloromethane and hydrogen chloride.

$$CH_4(g) + Cl_2(g) \rightarrow CH_3Cl(g) + HCl(g)$$

In a report of this reaction, a student came across a number of terms.

Illustrating your answer with an equation in **each** case, state what is meant by:

    (i)   homolytic fission,   *[2]*

    (ii) a propagation stage.   *[2]*

(b) One of the products of the reaction between ethane and chlorine is 1,1,1-trichloroethane.

The manufacture and use of 1,1,1-trichloroethane is now restricted because of its adverse effects on the ozone layer. However, the corresponding fluorocompound 1,1,1-trifluoroethane does not cause environmental problems in the ozone layer.

    (i)   Explain why only the chloro-compound has these adverse effects. *[2]*

    (ii) A sample of 1,1,1-trichloroethane is reacted with an excess of sodium hydroxide solution and then acidified.

        I One of the products of this reaction is liquid R whose mass spectrum shows a molecular ion at m/z 60.

          The infrared spectrum of R shows characteristic absorption frequencies at 1750 cm^{-1} and 2500-3500 cm^{-1}.

          Use this information, showing your working, to suggest a structural formula for liquid R.   *[4]*

        II Chloride ions are also produced when 1,1,1-trichloroethane reacts with aqueous sodium hydroxide. The products of the reaction are then acidified with nitric acid and the mixture tested for the presence of chloride ions.

          State the reagent(s) used and the observations when the mixture was tested for chloride ions.   *[2]*

Rhiannon's answer

(a) (i) Homolytic fission is when a bond breaks and each atom receives an electron. ✗ ①
$Cl-Cl \rightarrow 2Cl^{\bullet}$ ✓

(ii) A radical reacts and another is formed to carry on the reaction. ✓
$Cl^{\bullet} + CH_4 \rightarrow CH_3^{\bullet} + HCl$ ✓

(b) (i) The C–F bond is strong ✗
and is not broken in the ozone layer. ✗ ②

(ii) I $M_r = 60$ ✓
Infrared peaks show C=O ✗ O–H. ✗
Liquid R is ethanoic acid. ✗ ③

II The reagent used is aqueous silver nitrate ✓ and during the reaction a white colour is seen. ✗ ④

Examiner commentary

① Rhiannon does not score the first mark since two points were needed. The bond broken must be covalent and each of the joined atoms must receive one of the electrons.

② When a question includes words such as only/compare/ difference between, a comparison of some sort is required. In this case it was necessary to explain the difference in terms of the strengths of the carbon to halogen bond.

Rhiannon's answer does not include what causes the bond to break.

③ A mass spectrum is nearly always used to give the M_r and an infrared spectrum to show the bonds, and therefore functional groups, present. Rhiannon appreciates this but the absorption frequencies that indicate a particular bond must be specified. Rhiannon realises that R is ethanoic acid but the question states that a formula must be given.

④ Observations must include colour and if a solid precipitate is formed.

Rhiannon scores 5 out of 12.

David's answer

(a) (i) Homolytic fission is where a covalent bond breaks in an organic molecule. ✗ ①
$Cl-Cl \rightarrow Cl^{\bullet} + Cl^{\bullet}$ ✓

(ii) In a propagation stage, a radical takes part and regenerates another. ✓
$CH_3^{\bullet} + Cl_2 \rightarrow Cl^{\bullet} + CH_3Cl$ ✓ ②

(b) (i) Only the chlorocompound has adverse effects because the C–F bond is stronger than the C–Cl bond ✓ so that it is not broken by uv radiation. ✓

(ii) I The m/z peak quoted shows that the M_r is 60 ✓
In the IR spectrum the peak at 1750 cm^{-1} shows the presence of C=O ✓ and the one at 2500 to 3500 cm^{-1} shows O–H ✓
This means R is ethanoic acid CH_3COOH. ✓

II $AgNO_3$ is used ✓ and this gives a white precipitate. ✓ ③

Examiner commentary

① David's answer does not receive the first mark either for the same reasons as Rhiannon. Note that David used a different equation from Rhiannon but both are acceptable to show homolytic fission.

② Although different from Rhiannon's answer, both equations and definitions are acceptable.

③ Since the question says 'state the reagents', names or formulae are acceptable. When a question asks for an observation and a solid is formed, answers must specify this and its colour.

David scores 11 out of 12.

Alkenes

30 Compound A can be converted to 2-bromobut-2-ene in two steps:

$$H_3C, \quad CH_3 \xrightarrow[\text{Step 1}]{Br_2} \quad \begin{matrix} CH_3 & CH_3 \\ | & | \\ H-C-C-H \\ | & | \\ Br & Br \end{matrix} \xrightarrow[\text{Step 2}]{} \quad CH_3, \quad CH_3 \\ C=C \\ H \quad H \qquad \qquad \qquad \qquad \qquad \qquad H \quad Br$$

compound A   compound B

(a) During step 1, compound A is bubbled through bromine water to produce a layer of compound B.

    (i) Give the colour change that would be noted during step 1. [1]

    (ii) **Name** compound B. [1]

    (iii) Step 2 is performed using similar reagents and conditions to those used in the production of ethene from bromoethane. Give the reagents and conditions required for this reaction. [2]

(b) (i) Compound A also reacts with hydrogen bromide, HBr. Give the mechanism for this reaction. [4]

    (ii) Classify the mechanism of the reaction in (b) (i). [1]

Rhiannon's answer

(a) (i) It turns colourless. ✗ ①

    (ii) 2-dibromobutane. ✗ ②

    (iii) Sodium hydroxide ✓, dissolved in ethanol ✗ ③

(b) (i)

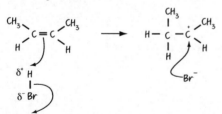

dipoles ✓ arrows ✗ structure of carbocation ✓ arrow from Br⁻ ✗ ④

    (ii) Addition ✗ ⑤

Examiner commentary

① When a question asks for the colour change, or observations, it is essential to include the colour at the start and end of the reaction.

② In the name of a disubstituted compound the positions of both substituents must be specified.

③ Many questions ask for reagents and the conditions needed for a reaction. Rhiannon states a correct reagent, an alkali, and recognises that it must be dissolved in ethanol. However, heating/reflux is an essential condition. This is true in many reactions in organic chemistry!

④ The dipoles and first arrow are correct.

However, the second arrow starts correctly from the bond but does not go to the bromine.

The structures of the carbocation are acceptable – the + is quite clearly on the correct carbon.

A curly arrow in a mechanism shows the movement of a pair of electrons. Rhiannon's final arrow does not start from a lone pair.

⑤ 'Addition' is not wrong but, in classifying a reaction, the nature of the initial attack should be stated (as well as the overall effect of the reaction).

Rhiannon scores 3 out of 9.

David's answer

(a) (i) Colour change is brown to colourless ✓ ①

(ii) 2,3-dibromobutane ✓

(iii) Sodium hydroxide ✓ dissolved in ethanol and then refluxed with compound B. ✓

(b) (i)

dipoles ✓ arrows ✓ structure of carbocation ✓ arrow from lone pair ✓ ②

(ii) Electrophilic addition ✓ ③

Examiner commentary

① The initial and final colours are quoted.

② It is really important that arrows start and finish in the correct places. They generally start at a bond or a lone pair.

③ The types of initial attack you will see in this unit are radical, electrophilic and nucleophilic.

David scores all 9 available marks.

Answers to Quickfire questions

1.1

1 (a) HCl (b) H_2SO_4 (c) NH_3 (d) CH_4.

2. Carbon 4 atoms, hydrogen 8 atoms, oxygen 4 atoms.

3 6 atoms.

4 (a) Na_2CO_3 (b) $BaSO_4$ (c) $(NH_4)_2SO_4$.

5 (a) -4 (b) $+5$ (c) $+3$.

6 (a) $Na_2CO_3 + 2HCl \rightarrow 2NaCl + H_2O + CO_2$

   (b) $3NO_2 + H_2O \rightarrow NO + 2HNO_3$

7 (a) $Mg(s) + 2H^+(aq) \rightarrow Mg^{2+}(aq) + H_2(g)$

   (b) $Pb^{2+}(aq) + 2I^-(aq) \rightarrow PbI_2(s)$

1.2

Extra 1

a) Zn-64: 30p, 34 n, 30e

   Zn-66: 30p, 36n, 30e.

b) (i) 8p, 10e (ii) 82p, 80e.

c) No difference because they have the same number of electrons in the outer shell.

1 Radioactive emissions are stopped by a lead shield thus preventing escape of the radioactivity.

2 ^{18}O.

3 21.2 years.

Extra 2

a) Radiation is ionising/releases high energy/ causes cell mutation.

   This causes radiation burns/radiation sickness/cancer.

b) α particles have low penetration depth and are able to be stopped by a few centimeters of air, or by the skin, therefore they cannot enter the body from outside. However, if ingested they cannot leave the body and since they are strongly ionising they can easily damage the DNA of a cell in the body and cause biological damage.

4 E.g., Carbon-14 used in radio-dating/potassium-40 used to estimate the geological age of rocks.

5 Some of the beta radiation is absorbed while passing through the product. If the product is made too thick or thin, a correspondingly different amount of radiation will be absorbed.

6 (a) $1s^2 2s^2 2p^6 3s^2 3p^6 3d^{10} 4s^1$

   (b) (i)

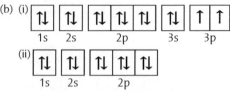

   (ii)

⇅	⇅	⇅	⇅	⇅
1s	2s		2p	

7 (a) Increase, because nuclear charge increases steadily but there is not much change in shielding.

   (b) Decrease, because outer electron has increased shielding from inner electrons and it is further from the nucleus.

8 $Mg^+(g) \rightarrow Mg^{2+}(g) + e^-$

9 Group 2 because there is a large jump in energy between the 2nd and 3rd ionisation energy, therefore the third electron has been removed from a new shell.

10 (a) Neon has the higher frequency since $f \propto 1/\lambda$.

   (b) Neon has the higher energy since $E \propto f$.

11 In absorption spectra, energy is absorbed from light causing electrons to move from a lower energy level to a higher one. It is seen as dark lines against a bright background.

   In emission spectra, energy is emitted as electrons fall back from a higher energy level to a lower one. It is seen as coloured lines against a black background.

12 Measuring the convergent frequency of the Lyman series (difference from $n = 1$ to $n = \infty$) and using $\Delta E = hf$ allows the ionisation energy to be calculated. The value of ΔE is multiplied by Avogadro's constant to give the first ionisation energy for a mole of atoms.

13 519 kJ.

1.3

1 (a) 331 (b) 246.5

Extra 1

a) To prevent the ions produced in the spectrometer from colliding with air molecules.

b) 87.7

c) $35.5 = \dfrac{(35 \times 75) + (x \times 25)}{100}$ therefore $3550 = 2625 + 25x$

   $25x = 925$ $x = 37$

2 The molecular ion splits to give Cl⁻ ions.

3 (a) 0.0465 (b) 74

4 VCl_2.

5 37.3 cm³.

6 20.7 cm³.

7 15.7 dm³.

8 10.6 g

9 6.15 mol dm⁻³.

Extra 2

a) 0.200 mol dm⁻³.

b) 37.0 %

c) 85.1 %

10 52.6 %

11 (a) 3 (b) 4 (c) 4

1.4

1

2 Cl–Cl< I–Cl ,H–Cl <Be–O

3 The bonds between the strongly bonded N_2 units are weak van der Waals forces of the induced dipole–induced dipole type.

4 Van der Waals < hydrogen < covalent

5

6

7 Tetrahedral, tetrahedral, trigonal bipyramid and linear respectively.

1.6

1 B^{3+} and C^{4+}. Ionisation energies are increasing across the period and it is energetically unfavourable to remove three or four electrons successively.

2 Across a period, e.g. between Li and F, electrons are being added to the same main shell so that there is little extra electron shielding to counteract the increased nuclear charge. However, the Na electron to be lost is screened by the full inner shell built up in the Li row and this offsets the extra protons in the nucleus.

3 (a) Na (0), Cl (0), Na⁺(+1) Cl⁻ (–1)

(b) –1

(c) Na (1), O(–2), S (6); N (5), F (–1); O(0); C (3), H (1)

Extra

a) 0

b) +6 +6

c) +2, +2.5

d) 0, +4

4 (a) Yellow Na, brick red Ca, apple green Ba and lilac K.

(b) (i) the hydroxides, (ii) the sulfates.

5 ½ Cl_2 (0) + K(1)Br(–1) = ½ Br_2(0) + K(1)Cl(–1)

1.7

1 Dynamic equilibrium is when the forward and reverse reactions occur at the same rate.

2 (a) The colour will become lighter. The position of equilibrium moves to the right because there are fewer gaseous moles on the right-hand side.

(b) The colour will become darker. Since the forward reaction is exothermic, the position of equilibrium moves to the left in the endothermic direction.

3 $K_c = \dfrac{[NH_3]^2}{[N_2][H_2]^3}$ dm⁶ mol⁻²

4 0.0243 mol dm⁻³.

5 (a) $MgO + H_2SO_4 \rightarrow MgSO_4 + H_2O$

(b) $CaCO_3 + 2HNO_3 \rightarrow Ca(NO_3)_2 + H_2O + CO_2$

6 A weak acid is one that partially dissociates in aqueous solution.

A dilute acid is one that contains a large quantity of water.

7 (a) 2.3 (b) 0.050 mol dm⁻³.

8 It reacts with atmospheric carbon dioxide and has a low molar mass.

2.1

Extra 1

a) (i) It's an element in its standard state. (ii) –1172 kJ mol⁻¹

b) –157 kJ mol⁻¹

1 –316 kJ mol⁻¹

2 –126 kJ mol⁻¹

3 –1251 kJ mol⁻¹

4 –52.7 kJ mol⁻¹

5 4.1 °C

2.2

1 (a) 3.16×10^{-3} mol dm^{-3} s^{-1}.

  (b) The initial rate of reaction would be greater because at the beginning of the reaction the concentrations of the reactants are greater therefore there is a greater chance that there will be more effective collisions.

2 (a)

  (b) 238 kJ mol^{-1}

3 Food spoilage is caused by bacteria. Low temperature in fridges reduces the activity of bacteria and stops them from growing freely.

4 Lower temperatures and pressures can be used so less CO_2 is formed during energy production.

They are biodegradeable, therefore they can be easily disposed.

5 Increase the concentration, temperature or pH of the hydrogen peroxide solution, add a catalyst such as manganese(IV) oxide.

6 Measure the change in pressure at various times using a manometer.

Measure the change in colour of NO_2 formation over time using a colorimeter.

7 An insoluble layer of $CaSO_4$ forms preventing the acid from reacting with the carbonate.

2.3

1 Chemical Equilibrium
Thermochemistry – energy and enthalpy
Kinetics – the rates of chemical reactions

2 The mass of the required product as a percentage of the total mass of the reactants.
Or using formulae this could be expressed as the ratio of the mass of the atoms used to the total mass of all the atoms in the reactants as a percentage.

Extra

a) + No carbon dioxide formed; – not present on Earth/ explosive gas.

b) + No organic solvent (VOC) used; – Used under pressure.

c) (i) + Low operating temp., no heating needed, no added solvent or co-product to be removed.

  (ii) – Higher temp and pressure needed, water co-product to be removed.

NB other answers possible

2.4

1 There are 6 carbon atoms in a continuous chain and therefore the name is based on hex.

2 (a)

  (b)

3 (a) 3-methyl pent-1-ene

  (b) 2-bromobutan-2,3-diol

4 (a) C_5H_9Cl

  (b)

  (c)

5 Pentanoic acid

6 $C_n H_{2n}$

7 $C_{72}H_{146}$

8 Ratio C : H : Br =
$$\frac{12.78}{12} : \frac{2.15}{1} : \frac{85.07}{80} =$$
1.07 : 2.15 : 1.06

Divide by smallest = 1 : 2 : 1

Empirical formula = CH_2Br

Empirical formula mass = 94 and actual molecular mass approximately 188.

Molecular formula = $C_2H_4Br_2$

9 Any 3 of:

hexane

2-methylpropane

3-methylpropane

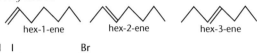

2,2-dimethylbutane   2,3-dimethylbutane

10 Any two of:

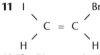

hex-1-ene   hex-2-ene   hex-3-ene

11

I   Br
 \\   /
   C = C
 /   \\
H   H

12 The E isomer. On C1 Br has a higher Ar than H and on C2 Cl has a higher Ar than C.

13 Any value in the range 80°C to 120°C

14

H H H H H   and CH_3
| | | | | |
H — C — C — C — C — C — H CH_3 — C — CH_3
| | | | | |
H H H H H CH_3

The more branching that is present the smaller is the surface area for Van der Waals forces. This means the boiling temperature is less.

2.5

1 (a) Sulfur dioxide and nitrogen dioxide

(b) Sulfur dioxide from burning fossil fuels and from volcanic activity.

Nitrogen dioxide from internal combustion engines and from lightning.

2 The black smoke is carbon. This is formed if the engine is not adjusted correctly and incomplete combustion occurs. This type of reaction is less exothermic than complete combustion.

3 $C_{90}H_{182}$

4 There are no double bonds, i.e. areas of high electron density and no dipoles.

5 Radicals are reactive because they have an unpaired electron and therefore want to gain another one.

6 Initiation $Cl_2 \rightarrow 2Cl\bullet$ Propagation $CH_4 + Cl\bullet \rightarrow CH_3\bullet + HCl$
Termination $2CH_3\bullet \rightarrow C_2H_6$

7 $CH_4 + 3Cl_2 \rightarrow CHCl_3 + 3HCl$

8 It is heterolytic bond fission because when the electrophile attacks and the bond breaks both the bonded electrons go to one of the atoms of the electrophile.

9 The melting temperature is raised and therefore they are more likely to be solids at room temperature.

10 $(CH_3)_2C=CH_2 + HBr \rightarrow (CH_3)_2CBrCH_3$

11 Poly(1-chloro, 2-cyanoethene)

12

⌐Cl CN Cl CN⌐
 | | | |
─C — C — C — C─
 | | | |
⌐H   H   H   H⌐

13 Empirical formula is CH_2

14

Cl   OH
 \\   /
   C = C
 /   \\
H   H

15 The melting temperature would be lower if the chain lengths in the polymer chain were shorter and if the polymer contained branched chains.

2.6

1

Cl Cl H H
| | | |
H — C — C — C — C — H
| | | |
H H H H

2

(structure)

3 It is not a satisfactory method because it has a radical mechanism and this is rather indiscriminate. Polysubstitution can occur.

4 Heat with aqueous sodium hydroxide.

Neutralise the excess sodium hydroxide with dilute nitric acid.

Add aqueous silver nitrate.

The expected result is a yellow precipitate that is insoluble in aqueous ammonia.

5 $Ag^+(aq) + Cl^-(aq) \rightarrow AgCl(s)$

6 2- Chlorobutane can lose the Cl and the H on either side of the Cl to form but-1-ene and but-2-ene. In 3-chloropentane the Cl is in the middle so whichever H is lost the only possible product is pent-2-ene.

7 Halogenoalknes have a non-polar section in their structure. Grease also is largely non-polar and so grease dissolves in the halogenoalkane.

8 CFCs can be liquefied by being put under pressure.

9 In a propagation step a radical is present before the reaction and a radical is present after the reaction.

10 $2O_3 \rightarrow 3O_2$

11 HFCs are being increasingly used because the C–F bond is stronger than the C–Cl bond. This means that F • radicals are not generally formed.

2.7

1 The reaction is exothermic so a high temperature sends the equilibrium to the left and lowers the yield of ethanol.

2 Fermentation is an enzyme-catalysed reaction. Enzymes are denatured/destroyed if the temperature is too high – a temperature around the normal human body temperature is usually used.

3 Ethanol can be removed by fractional distillation.

4 Carbon dioxide is generally considered to cause global warming. The combustion of biofuels does produce carbon dioxide but biofuels are derived from plant material that took in carbon dioxide when it was growing. They are on this basis carbon neutral.

5 (a) Primary

(b) Tertiary

(c) Secondary

(d) Primary

6 (a)

$$H-\overset{\overset{\displaystyle H}{|}}{\underset{\underset{\displaystyle H}{|}}{C}}-\overset{\overset{\displaystyle H}{|}}{\underset{\underset{\displaystyle H}{|}}{C}}-\overset{\overset{\displaystyle H}{|}}{\underset{\underset{\displaystyle H}{|}}{C}}-OH \rightarrow \overset{\overset{\displaystyle H}{\diagdown}}{\underset{\underset{\displaystyle H}{\diagup}}{C}}=\overset{\overset{\displaystyle H}{|}}{C}-\overset{\overset{\displaystyle H}{|}}{\underset{\underset{\displaystyle H}{|}}{C}}-H + H_2O$$

(b)

$$H-\overset{\overset{\displaystyle H}{|}}{\underset{\underset{\displaystyle H}{|}}{C}}-\overset{\overset{\displaystyle H}{|}}{\underset{\underset{\displaystyle H}{|}}{C}}-\overset{\overset{\displaystyle H}{|}}{\underset{\underset{\displaystyle H}{|}}{C}}-OH + 2\,[O] \rightarrow H-\overset{\overset{\displaystyle H}{|}}{\underset{\underset{\displaystyle H}{|}}{C}}-\overset{\overset{\displaystyle H}{|}}{\underset{\underset{\displaystyle H}{|}}{C}}-\overset{\overset{\displaystyle O}{\diagup\diagup}}{\underset{\underset{\displaystyle OH}{\diagdown}}{C}} + H_2O$$

7 Tertiary alcohols are not normally oxidised as they have no suitable H that can be lost.

8

$$-C\overset{\displaystyle \diagup\diagup O}{\diagdown H}$$

9 The colour changes from orange to green.

10 (a) CH_3COONa

(b) $(CH_3CH_2COO)_2Zn$

11

$$H-\overset{\overset{\displaystyle CH_3}{|}}{\underset{\underset{\displaystyle C}{|}}{C}}-O-\overset{\overset{\displaystyle O}{||}}{C}-CH_3$$
$$H-\overset{|}{\underset{|}{C}}-H$$
$$\underset{\displaystyle CH_3}{|}$$

12 Carboxylic acids are **weak acids** because they are **partially** ionised. They react with alkalis to form **salts**.

2.8

1 Spectroscopic techniques need less sample.

2 The y-axis gives the abundance of the fragments. This is not generally very useful in finding the structure of the molecule.

3 m/z at 46 – this is the M_r of ethanol.

4 (a) The peak at 78 is due to $C_3H_7{}^{35}Cl$ and that at 80 is due to $C_3H_7{}^{37}Cl$.

(b) The peak at 78 would be 3 times as large as that at 80 since there is 3 times more ^{35}Cl than ^{37}Cl.

5 At m/z 74.

CH_3 has been lost from butan-1-ol to give the peak at 59.

6 The C–H bond causes absorptions in this range and, since this is present in nearly all organic compounds, peaks here do not give useful information.

7

$$CH_3C\overset{\displaystyle \diagup\diagup O}{\diagdown OCH_3}$$

This absorption shows the presence of C=O. This is present in carbonyl compounds, carboxylic acids and esters.

8 (a) Peak heights in ^{13}C NMR spectra are not generally useful.

(b) The peak heights in a ^1H NMR spectrum tell you the number of hydrogen atoms present in each environment.

Index